Jan. 15, 2015

Roy,

Mary asked me to send you a
copy of the book I co-authored.

If you are suffering from
insomnia this book is guaranteed
to cure you. It's the best thing
on the market.

With love & affection,
Frank

MANAGING PROFESSIONAL SERVICE DELIVERY

9 RULES FOR SUCCESS

MANAGING PROFESSIONAL SERVICE DELIVERY

9 RULES FOR SUCCESS

Barry M. Mundt • Francis J. Smith • Stephen D. Egan Jr.

CRC Press
Taylor & Francis Group
Boca Raton   London   New York

CRC Press is an imprint of the
Taylor & Francis Group, an **informa** business

CRC Press
Taylor & Francis Group
6000 Broken Sound Parkway NW, Suite 300
Boca Raton, FL 33487-2742

© 2014 by Taylor & Francis Group, LLC
CRC Press is an imprint of Taylor & Francis Group, an Informa business

No claim to original U.S. Government works

Printed on acid-free paper
Version Date: 20140418

International Standard Book Number-13: 978-1-4398-5142-5 (Hardback)

Visit the Taylor & Francis Web site at
http://www.taylorandfrancis.com

and the CRC Press Web site at
http://www.crcpress.com

Contents

Preface: 9 Rules for Success

The goal of this book is to help you understand engagement-tested methods for success at every step in delivering a professional service. Large professional service firms have the resources to develop practice guides for their staff members; however, sole proprietors and small- to medium-sized firms typically do not. We hope this book guides you—starting with the 9 Rules for Success—through the maze of delivering your professional service.

For those of you who are unable to exhibit internal discipline and control (see Rule 2, below), we provide the 9 Rules for Success up front, where you can find them easily. We also indicate in which chapter(s) each of the 9 Rules is discussed, should you wish to explore further. Following are the 9 Rules:

Rule 1. Clearly define your market niche (industry or industries, geographical coverage, client size, and list of services) to create a unique and powerful offering to potential clients [Chapter 1].

Rule 2. Develop and implement a system and culture of internal discipline and control to ensure consistency of service, efficiency of operation, and quality and reliability of deliverables. Then train, mentor, and monitor personnel regarding engagement management policies and procedures [Chapter 2].

Rule 3. Establish and enforce engagement documentation standards, including those for proposals, progress reports, and deliverables [Chapters 2, 3, 4, 6, and 7].

Rule 4. Practice what you preach regarding internal culture, policies, procedures, and standards [Chapter 2].

Rule 5. Prepare complete and definitive service proposals, contracts, and engagement work plans that evaluate and accommodate engagement risks for both the provider and the client, so the client knows what can be expected in terms of scope, work plan, schedule, deliverables, and cost [Chapters 3 and 4].

Rule 6. Be flexible and adaptive to the "real world" once the project starts, to manage the dynamic between client expectations and what's *really* happening within both the firm and the client [Chapters 5 and 6].

Rule 7. Implement a firm–client communication plan that will ensure clear and frequent discussion of engagement progress and status [Chapters 4 and 6].

Rule 8. Bill and collect frequently. This will improve cash flow and alert you sooner if the engagement is in trouble with the client [Chapters 6 and 7].

Rule 9. Conduct independent status/quality reviews of the engagement while it is in process and subsequent to conclusion. Involve key client contacts in the reviews [Chapters 6 and 7].

DELIVERY OF PROFESSIONAL SERVICES

This book is designed to provide service professionals with the basics of *how* to deliver service to their clients; it does *not* deal with *what* services the professional actually delivers. Although the nature of the service delivered varies significantly from profession to profession, the *way* the service is delivered tends to be fairly consistent among professions—or, at least, it should be.

PURPOSE OF THIS BOOK

This book has several specific purposes:

- Provide a structure for how to manage professional service delivery, from start to finish
- Give tips on how to set up an environment and develop a culture that will result in superior service delivery—such that the delivery process incorporates rigorous internal discipline and control
- Discuss rapid implementation and deployment concepts that can be attained without compromising internal discipline and control
- Provide examples of documentation standards for service proposals and deliverables (reports)
- Review the application of internal discipline and control in two projects conducted by the authors

Managing Professional Service Delivery should be considered as a primer or basic text, in the sense that it applies to a sole practitioner or to a firm that has, say, up to seventy-five practitioners and support staff—a small- to medium-sized firm. Much larger firms that are experiencing sustainability issues, perhaps related to the markets served or a loss of internal discipline and control, also should consider reviewing this book.

PROFESSIONAL SERVICE DEFINITIONS

The professional services "industry" is divided into a number of different sectors. The terminology used to describe various elements of the delivery process can vary significantly from sector to sector. To avoid confusion, we need to settle upon and define a single set of basic terminologies to be used in this book. In this section, we define the following terms, which will be used throughout the book:

- Professional service organizations
- Firm
- Client
- Engagement
- Project
- Deliverable
- Internal discipline and control

PROFESSIONAL SERVICE ORGANIZATIONS

Professional service organizations now span a wide range of disciplines. They often are thought to provide ongoing advice or perform customized work for the typically infrequent buyer of the service. The organization itself can be in the form of a partnership, a corporation, or an individual or sole practitioner. In the United States alone there are approximately 760,000 professional service firms, with total revenues on the order of $1.3 trillion.* Examples of professional service sectors include the following:

- Accounting and auditing
- Actuarial
- Advertising
- Appraisal
- Architecture
- Engineering
- Information technology

- Legal
- Management consulting
- Marketing
- Public adjusting
- Public relations
- Recruiting

This is by no means a complete list; however, it does serve to illustrate the wide breadth of the professional service industry. And it encompasses the types of professions for which this book has been written.

THE FIRM

We use the term *firm* throughout this book to refer to any professional service organization that provides service to a client.

THE CLIENT

The buyer of a professional service is called the client. Merriam Webster's online dictionary defines a client as "a person who engages the professional advice or services of another (such as a lawyer's client)."

Normally the client is external to the professional service firm. However, the client could be internal to the organization, such as a human resources, information technology, consulting, or engineering unit that provides services to other departments (for example, a human resources department may service operating departments by recruiting for open positions). In such cases, many of the service delivery management processes and activities discussed in this book will apply.

Often the client will seek out the professional service firm based on its reputation or due to a referral from a satisfied customer. Or the firm might seek out the potential client (or multiple clients in an industry), based on a specific marketing approach.

THE ENGAGEMENT

An engagement is *firm-oriented* in that the firm carries it out. It is characterized by a formal agreement between the professional service firm and the client. The nature

---

* http://selectusa.commerce.gov/industry-snapshots/professional-services-industry-united-states

of the engagement is described in the form of a contract between the firm and the client. Typically, the contract sets out the purpose of the engagement, the results to be delivered over a specified time period, and the compensation that the firm is expected to receive for the service.

THE PROJECT

A project is *client-oriented* to the extent that the client is willing to invest resources in it. Three key elements characterize a client's project:

- It is a unique venture with a defined beginning and end.
- It is carried out by people to meet a specific objective or set of objectives.
- It is defined within the parameters of scope, schedule, cost, and quality.

THE DELIVERABLE

A deliverable is any contractually required progress or end product that is provided by the firm to the client. For example, an architect's sketches and drawings might constitute deliverables, or a management consultant's progress reports, interim reports, and final report, as well.

INTERNAL DISCIPLINE AND CONTROL

Internal discipline and control encompass a business philosophy, an organization structure, and an operating methodology (including policies, procedures, templates, and forms) that allow a professional service firm to operate both *effectively* and *efficiently*—thereby yielding *productivity* that enhances the *quality* of its work and the probability of successful service delivery. It requires that all "frontline" service delivery and "backroom" support personnel adopt the philosophy and abide by the established policies, procedures, and methodologies to ensure that the firm makes a fair return on its engagements and that the client is satisfied with the results. Some major professional service businesses and engagements have failed in recent years, due primarily to the loss of internal discipline and control regarding the engagement process and/or deliverables.

WHAT CONSTITUTES A SUCCESSFUL ENGAGEMENT?

Both the client and the service delivery firm have a set of goals for an engagement, some of which are in common. From the client's perspective, the goals are to receive the desired result (deliverables) on time and within budget, with the least disruption of its operations. The service delivery firm should have the same goals, but with the addition of making a fair return *and* developing a satisfied client. Of course, if the client gets the desired result, on time, within budget, and with little disruption, they will be satisfied. In the end, achieving both the client's and the service deliverer's goals will require that the engagement contract be specific and complete, and that both parties do what they said they would do.

ABOUT THIS BOOK

The structure of *Managing Professional Service Delivery: 9 Rules for Success* closely follows the life cycle of a professional service engagement. At or near the conclusion of each chapter and the Preface are some related tips for sole proprietors and small firms.

We start before you even consider your first engagement opportunity by outlining a context, in Chapter 1, about what you and your professional service firm should already have done to define your niche in the marketplace. We discuss formational decisions, such as mission, vision, values, target industries and clients, and services—which in sum is your niche.

In Chapter 2, we discuss how to establish a philosophy and culture of internal discipline and control (IDC), as well as how to document and communicate the related IDC methodology (including policies, procedures, and supporting technology, tools, templates, and forms). Finally, we provide some tips on how to organize, hire, and train the professional staff with regard to the IDC philosophy and methodology.

Chapter 3 deals with selling an engagement (or what some professional service organizations would call "business development"). This starts with defining the services to be delivered, and identifying and contacting the potential clients. When a party shows interest, the next step is to prepare and present the service proposal, which should result in a closed sale and the execution of a specific contract.

Setting up the engagement (Chapter 4) involves more than just feeding the relevant data into the firm's engagement control system. Perhaps the most important part of the engagement occurs during this phase: assignment of the engagement team, particularly the engagement director and project manager—people who have the right technical skills and personalities to work well with key client contacts. They then will develop the detailed work plan, schedule, and budget—as well as a risk mitigation strategy and plan. The communication strategy and plan, both internally and with the client, become a part of the work plan and schedule.

In Chapter 5, we discuss some situations that may require adaptation of the normal engagement plan, such as

- Dealing with an internal client, rather than an external entity
- Accommodating cultural differences between the service provider and the client
- Incorporating a new service paradigm: Rapid Change and Rapid Deployment

Chapter 6 gets to the heart of the matter—conducting the engagement. Here we deal with the nuts and bolts of getting the engagement underway, and then delivering the results on time and within budget. This includes managing the engagement documentation; managing the work plan, schedule, and budget; engagement management and supervision; progress monitoring and reporting (internally and to the client); preparation of the deliverables; and progress billing and collection. At times, there may be a need to amend the scope of work, the work plan and schedule, and/or the contract. Additionally, there may be a desire to conduct an independent review of the engagement while it is in process.

Chapter 7 deals with closing the engagement. This involves preparing and submitting a final report or letter to the client that summarizes what has been accomplished—typically accompanied by a final invoice. Often there is a postcompletion evaluation of the engagement, the results of which may be shared with the client. A follow-up meeting with the client within, say, six months may be useful to maintain the relationship and determine if there is a need to provide further assistance.

In Chapter 8, we reiterate the 9 Rules. Then we provide two case studies of management consulting engagements that the authors conducted, which illustrate application of the 9 Rules.

In the Appendixes, we provide a Proposal Preparation Guide (Appendix A) and a Report (Deliverable) Preparation Guide (Appendix B).

TIPS FOR SOLE PROPRIETORS AND SMALL FIRMS

If you left a company or a public or nonprofit organization to join or create a professional service firm, welcome to a new world in which you need a client to get paid. No clients means no revenue—even as the bills for your business expenses keep coming. This change will be a major paradigm shift from the corporate or government life to the professional service model. Following are some tips for sole proprietors and small firms:

1. You now are working on engagements (for the firm) and not projects (for the company, government, or nonprofit).
2. Instead of an hourly, weekly, or monthly pay rate (and paycheck), you have an hourly or daily billing rate that is two to four times your cost of service. Get used to saying something like, "My hourly billing rate is $150," without blushing, choking, or breaking out in a sweat.
3. The client decides if you are successful, not you or even your boss.
4. The "beach" is a bad place to be, as it means you don't have a billable engagement and your time as a professional service provider is running out. Like "sand through the hourglass", these are the (unbilled) days of your (soon to be short work) life.
5. As a sole proprietor, you need a system of internal discipline and control, even if you are the only one being disciplined and controlled. So, develop an IDC on the fly or from scratch, or borrow it, but put it in place somehow if you want to be successful.
6. Welcome to Barry's, Frank's, and Steve's world … it's fun and rewarding, but stressful at times.
7. This book is for *you*!

About the Authors

Barry Mundt and Frank Smith have had both a personal and a business relationship for more than twenty years. They co-authored a chapter in Gavriel Salvendy's *Handbook of Industrial Engineering, Third Edition*,[1] entitled "Managing Professional Services Projects" (Barry also co-authored another chapter, "Enterprise Concept: Business Modeling Analysis and Design").

Barry and Steve Egan's relationship goes back to the early 1980s, when Barry hired Steve to be a part of his management consulting practice in the Atlanta office of Peat, Marwick, Mitchell & Co. (now KPMG LLP). Barry is Steve's mentor as a consultant and a long-time advisor to Steve's consulting practice.

BARRY MUNDT

Barry earned a bachelor of science degree in industrial engineering from Stanford University and a master of business administration degree from Santa Clara University. He began his professional career in the aerospace industry, first with Aerojet General Corp. and then with Lockheed Missiles and Space Company. He joined a consulting arm of C-E-I-R, Inc. (a large computer service firm) in 1961, which was headed by three of his former Stanford professors. Barry has been a consultant to private- and public-sector organizations since that time.

In 1965, Barry was asked to join the management consulting department of what was then known as Peat, Marwick, Mitchell & Co. (PMM), a global public accounting, tax advisory, and management consulting partnership. He was appointed partner in 1973. When Peat Marwick International merged with Klynveld Main Goerdeler in 1987 to form KPMG LLP, Barry assumed the role of international management consulting coordinating partner in Amsterdam, the Netherlands. Barry was assigned to the firm's Department of Professional Practice for the two years prior to his retirement. During that time, he took the lead in preparing KPMG's Performance Improvement Consulting Services Manual.

When Barry retired from KPMG in 1995, after thirty years of service, he set up his own management consulting practice, specializing in services to professional service firms. In that capacity, he has written KPMG's Global Management Consulting and Advisory Services Engagement Conduct Guides and participated in the preparation of Adjusters International's suite of Disaster Recovery Consulting Operational Guides, which includes an Engagement Conduct Guide.

Barry is a fellow, life member, and past president of the Institute of Industrial Engineers. He is a member of the board of directors of Adjusters International, Inc., a nationwide group of public insurance adjusters and disaster recovery consultants.

FRANK SMITH

Frank earned a bachelor of science degree in education from the University of Connecticut and a master of science degree in management from Rensselaer

Polytechnic Institute. He began his professional career teaching in the Pomperaug Regional School District #15 in Connecticut. He later went on to work in several high-technology, aerospace, and defense companies in line management positions. In 1987, he joined Peat, Marwick, Mitchell & Co. (which soon became KPMG), where he served in the Technology and Operations group of KPMG Management Consulting. Frank left KPMG in 1992 as a senior manager and set up his own management consulting practice. His services focus on delivery of performance improvements to professional service, health care, and manufacturing organizations.

STEVE EGAN

Steve earned a bachelor of arts degree in history and government from Norwich University and a master of arts degree in government from American University in Washington, DC. Steve's first job was as a research assistant for the Governor's Crime Commission in Vermont, followed by eight years as a management and budget analyst in the finance department of Fulton County, Georgia, where he also served as an internal management consultant at times.

In 1981 Steve joined PMM's government consulting practice and, in 1987, moved to David M. Griffith and Associates. In 1992 he formed ECA Strategies, a sole proprietorship focused on management and financial services to state and local governments. Since 1995, ECA Strategies has been affiliated with the Mercer Group, Inc., a national public-sector consulting firm specializing in strategic and functional business planning, management and organizational improvement, human resource management, and executive recruitment.

In thirty-two years of consulting, Steve has conducted more than 160 planning, management, operations, and financial studies and is a nationally known expert in strategic and functional business planning, service delivery alternatives, and organizational development. He is also known for his work in management and organizational improvement for public works, utilities, parks and recreation, planning and community development, and public safety operations.

From 2002 to 2005 he served as the interim Public Services and Water Director for the City of Highland Park, Michigan, which at the time was a fiscally distressed city run by an emergency financial manager appointed by the State of Michigan.

ENDNOTES

1. Salvendy, Gavriel. 2001. *Handbook of Industrial Engineering: Technology and Operations Management*, Third Edition. New York: John Wiley & Sons, Inc.

1 Creating Your Niche in the Market

Rules for Creating Your Niche in the Market

Rule 1: Clearly define your market niche (industry or industries, geographical coverage, client size, and list of services) to create a unique and powerful offering to potential clients.

WHAT DRIVES YOUR WORK?

Creation or management of a professional service firm is the culmination of years of education, training, and work experience. It may involve just you ... or multiple owners or partners. Often the firm grows out of a personal conviction that you can serve the world better as an adviser to organizations rather than by working in one. Or, you just got laid off from your company, government agency, nonprofit, or professional service firm and are in a panic over what to do next to feed your family, instantly making you a one-person firm without a name, identity service, or income stream. Or, your firm creates itself because a potential client asks, "Can you help us?" or because you suggest, "Maybe I can help you."

Thomas Moore, author of *Care of the Soul* and *A Life at Work*, compares work to the alchemist's process of creating a finished product by refining raw materials. He says, "Anyone's quest to find their life work [in this case delivering professional services and building a professional services firm] is deep and mysterious ... involves changes and developments...requires patience, good powers of reflection and observation, and the courage to keep going when it seems nothing of worth is happening."[1]

Each professional service company will have a unique driving force in its creation, services, and operation. Paul Hawken, founder of Smith and Hawken, started his first natural food store in Boston to restore his health (due to asthma) by finding great, local sources of natural foods. Tom Chappell, the co-founder of Tom's of Maine, started his business because he cared about what is actually in the things we buy. A friend of one of our authors left a large CPA and consulting firm to join his step-father's surveying and engineering firm because he felt a "call" to work there—partly in response to a desire to balance his work life with a new wife and growing family.

FOUNDATIONAL DECISIONS

No matter how the firm started, the type of firm, the services provided, or the legal structure, the firm should be created and operated on the principles of internal discipline and control discussed in Chapter 2 of this book, as exemplified by the 9 Rules for Success. These principles are as important to a sole proprietor or small firm as to a large national or international corporation or a multipartner firm that has a regional focus. In every case, your goals in creating or joining a firm must include providing excellent services to clients and sustaining yourself and the firm over a lifetime of work—and perhaps into future generations of owners and partners.

Before you can apply the principles of a culture of internal discipline and control to products and services, you have to make foundational decisions regarding the focus, scope, features, and attributes of your firm in order to answer several important questions:

- What skills, experience, and ideas do I, and perhaps partners, bring to the business that will appeal to potential clients?
- What business should I/we be in now and in the future?
- How should the business be organized (e.g., corporation, LLC, partnership, sole proprietorship)?
- What resources are required to serve potential clients (e.g., people, offices, technology, methodologies)?
- What is the market for my services (e.g., industries, geographic area, type and size of client)?
- What is my unique niche within that market (e.g., service, clients, approach)?
- How do I make the business viable in the marketplace for the long term?
- If an existing business, do I need to reinvent the business now or in the future in order to remain viable, and how do I do that?

Some level of strategic business planning is the springboard to answering these questions and to starting or reinventing your business.

SHAPING STRATEGIC SUCCESS

In their classic book *Shaping Strategic Planning: Frogs, Dragons, Bees, and Turkey Tails*, William Pfeiffer, Leonard Goodstein, and Timothy Nolan[2] identify the following keys for shaping strategic success in a private-sector organization, then fill up the book expounding on each key:

- Base decisions on values
- Make the mission crystal clear
- Sound a rallying cry (leadership!)
- Persevere
- Promote and reward risk-taking
- Encourage innovation

- Monitor and manage "down board" (like a chess player who is thinking several moves ahead)
- Maintain a market (and client) focus

James L. Mercer, president and CEO of the Mercer Group, Inc., says the objective of a strategic planning process for governments and nonprofits (which holds for a professional service firm with some adjustments) is to "increase organizational performance through examination of [market] needs, establishment of organizational goals, and identification of steps necessary to achieve these goals. Strategic planning concerns itself with establishing major directions for the firm (such as its purpose/mission), major clients to serve, major problems [or issues] to pursue, and major service delivery approaches."[3]

TWO STRATEGIC PLANNING MODELS

Pfeiffer, Goodstein, and Nolan offer an Applied Strategic Planning Model to incorporate their keys for strategic success into a planning process.[4] The Mercer Group, Inc. uses a similar strategic planning model to guide state and local governments, nonprofit organizations, and the firm in defining and maintaining an effective strategic direction. The model was developed by Jim Mercer for use in his strategic planning work; it has been applied more than 100 times on client engagements.

The similarities in these two strategic planning models, one for the private sector and one for the public sector, are striking:

- Preparing and planning to plan are critical first steps.
- Mission, vision, and values are defined early in the process and drive the development and implementation of strategies, initiatives, goals, and objectives.
- The environmental scan and continuous monitoring of the environment (industry or sector events and trends, potential clients, and active competitors in particular) provide crucial insights into opportunities, challenges, and threats to achieving the mission.
- Internal assessments and gap analyses show capabilities and shortfalls in implementing the plan.
- Tactical/operations plans are integrated with higher-level strategies and goals to ensure alignment from top-to-bottom.
- Contingency plans (like risk mitigation strategies and plans) prepare the firm to respond to known and unknown challenges, issues, and events, such as the loss of a key client or a national recession.
- Critical success factors and performance indicators provide a road map to success and a scorecard for measuring success.
- Implementation is an ongoing process that provides feedback and encourages updates to the planning process and the plan.

The two models vary only in the sequencing of two key tasks—the *environmental scan* and *mission, vision, and values development*. Mercer puts the environmental scan before the development of mission, vision, and values—which makes sense in plans to serve long-standing and historically slow-changing governmental and nonprofit organizations. Pfeiffer, Goodstein, and Nolan reverse these tasks, with the development of mission, vision, and values coming earlier in their planning process for private firms. In either model, these steps inform each other and are modified as more information or experience is gained. For example, in one of our author's management consultations, periodic local government financial stress (an environmental factor) opens up revenue enhancement and productivity work (tactical opportunities).

ELEMENTS IN DEFINING A STRATEGIC DIRECTION

We've shown you two fairly comprehensive strategic planning models, one oriented to the private sector and the other to the public and nonprofit sectors. In contrast, Dilbert advises, "There are two major steps in building a business plan: (1) Gather information and (2) Ignore it."[5] Your choice of an approach to defining strategic direction is somewhere between these two alternatives (plan or not plan), but hopefully not too close to Dilbert's.

Key questions to be asked when creating or updating a strategic plan for a professional service firm include the following:

- Have we adequately *scanned the environment*, including industry trends, competitors, and other external factors impacting business success, and developed and acted on a SWOT (strengths, weaknesses, opportunities, threats) assessment? Are we continually scanning to keep ahead of awful surprises?
- Have we developed a clear *purpose/mission statement* for the firm and its component units, and is this statement up-to-date for today's opportunities and challenges?
- Is our *vision statement* clearly defined and are we looking out far enough … or too far?
- Have we adequately stated and communicated, as well as lived, our *core values*?
- Are *tactical or functional plans* in place for major business units (if you have them), including a *contingency plan* in case the "wheels fall off"?
- Do our *performance and financial plans* link to and express our mission, vision, values, and key goals and objectives?

Each of these questions implies a step in preparing a strategic plan for your business. So let's walk through each of the steps in the strategic planning process and apply them to a professional service firm.

Preparing to Plan

Before you embark on the planning process or update your plan, you should figure out how much of a plan you need. Steve Egan, like Dilbert in his Step 1, spent a lot of time in 1992 on strategic planning when he opened his own management consulting firm, then promptly decided a lot of what he developed was junk and threw it away, like Dilbert in his Step 2. His second approach to planning was to measure a few key business-sustaining activities (e.g., at least five business development contacts, one proposal, and eighty billable hours per month) and a few key indicators of success (e.g., win four to six projects each year, provide a minimum monthly cash flow, and add five to ten new references each year).

As a sole practitioner in the early years, Steve had to do the planning, marketing, managing, and conducting the management consulting engagements himself, with little help. He needed enough of a plan to make the marketing pay off and the engagements succeed, without wasting time or being overwhelmed. A philosophy and culture of internal discipline and control were incorporated into the planning process informally, but based on a great model learned at KPMG LLP, to ensure projects succeeded and clients became enthusiastic references.

The following sections describe the key elements of a useful strategic planning process.

ENVIRONMENTAL SCAN

The environmental scan assesses external trends and factors influencing the organization; opportunities, challenges, and threats created by the external environment; and the internal strengths and weaknesses that are available to respond appropriately or are necessary to correct to be successful. This scan also may be called a *situation analysis*, a *size-up*, an *environmental analysis*, or a strengths/weaknesses/opportunities/threats (*SWOT) analysis*. Regardless, the environmental scan also will identify the following:

- *Critical success factors*, which are the crucial elements to success in carrying out the mission (e.g., gaining a planned number of new clients each year)
- *Driving forces*, which are the focus of the organization as it attempts to be successful in carrying out its mission or achieving its vision (e.g., your passion for a few services and your methodology for delivering them)
- *Scenarios*, the overall strategy the organization plans to follow (e.g., serving sector clients in a set size or revenue range)

The environmental scan may include a more detailed assessment of the adequacy of human and tangible resources, policies, procedures, and operating practices to achieve the mission. These elements are critical to an effective internal control process.

The scanning process will be more formal in larger firms, perhaps with resources (such as a planning group) dedicated to analyzing external influences and measuring internal capabilities, then resolving shortfalls. This group may conduct a performance audit or internal assessment that develops a gap analysis, which identifies where the organization is lacking in order to respond to the market, specific industries, and service opportunities.

MISSION, VISION, AND VALUES

Once you understand your environment and opportunities, the planning process should define three key statements:

- *Mission Statement*—A statement of an organization's purpose or reason for being, which answers the questions, Why do we exist? What are we here to do together? What business are we in? What business *should* we be in? What business should we *not* be in? For example:
  - "To make our clients proud that they engaged us to provide management consulting services for them."
  - "Our purpose is to turn knowledge into value for the benefit of our clients, our people, and the capital markets."
- *Vision Statement*—What we want our business or organization to be at a designated future time; a picture of the future we want to create. A vision is a statement of where we want to go and what we will be like when we get there. For example:
  - "We will be the leading surveying and engineering firm for planned communities in our state."
  - "We will be the dominant legal services provider to cities, counties, and other units of local government in the state."
- *Core Values Statement*—An enduring belief that one set of behaviors, courses of action, or end results is preferable and will help our people work together organically and ethically to accomplish the vision; how we intend to operate as we pursue our vision; a guiding symbol of our behavior. Values drive an organization's culture and relationships. Core values also are called *guiding principles*. Some examples of organizational core values are as follows:
  - High-quality and cost-effective services
  - Creativity and innovation in providing services
  - Honesty and integrity in all we do
  - Courtesy and respect to each other and our clients
  - Open, two-way communication
  - Empower our employees and teams
  - To value diversity
  - Invest in our workforce
  - Be a great place to work

These three statements—vision, mission, and core values—define your purpose, your destination, and your conduct along the way.

TACTICAL OR FUNCTIONAL PLANS

Finally, strategic initiatives, goals, objectives, and tactical plans are developed to implement the mission and vision that have been based on the environmental scan and gap analysis. Following are definitions and examples of each term:

- *Goal*—General and timeless statements of a desired end result or outcome. A goal statement is a fairly broad definition of expectations and success. Goals are stated in terms of fundamental elements that support achievement of the vision. Goal attainment can be measured by "Yes, we did accomplish the goal," or "No, we didn't." For example: *Develop a practice base of $2 million per year in federal agency work.*

- *Strategic Initiative*—Provides direction on *what* to do to carry out the vision, mission, and goals. This represents a "road map" toward the end vision. For example: *Open a regional office in Washington, DC, to expand into the federal government market.*
- *Objective*—A tactical outcome or output statement that is specific, measurable, and has a time frame for accomplishment. Objectives are what people commit themselves to in support of achieving goals. SMART objectives are
  - **S**pecific
  - **M**easurable
  - **A**ssignable
  - **R**ealistic
  - **T**ime Related

These objectives measure the who, what, when, where, and how of reaching goals. For example: *Hire a Washington office partner and have the office open and running by the end of the current fiscal year.*

- *Action or Tactical Plans*—Action, tactical, functional, or operations plans typically are short to mid-range in time and arranged by organizational units or functions. They are intended to carry out the strategies, initiatives, goals, and objectives in the organizationwide strategic plan. For example, tactical plans might include the following:
  - Functional business plans for specific programs and departments
  - Operating and capital budgets
  - Information technology plan
  - Training and employee development plan
  - Compensation and benefits plans
  - Contingency plan for unforeseen circumstances (like the loss of that *big* client or a market slowdown)

Tactical plans for a very large firm would define operating and capital resource needs and functional business plans for each practice area.

Objectives and action plans, therefore, are the stepping-stones along the organization's path to accomplish goals and complete strategic initiatives.

IMPLEMENTATION AND FEEDBACK

Assuming you have worked in the industry you plan to service, you likely have a good feel for the environment—both in the early years of the business and as you adapt to changing situations. However, continued investment in scanning and analyzing the environment, particularly with regard to changing client needs and competitor capabilities, is critical to keeping the strategy fresh and relevant.

SAMPLE PROFESSIONAL SERVICE STRATEGIC PLAN

The key elements of Steve Egan's plan for his sole proprietorship, ECA Strategies, are shown below. The Mercer Group also has a strategic plan, within which the ECA plan is a component.

- *Mission*: Improve the performance and cost-effectiveness of state and local governments through analysis, creative ideas, and encouragement.
- *Vision*: Create a sustainable business that will last as long as I want to work and provide adequate resources to retire.
- *Values*: Do everything with integrity and honesty; create an atmosphere that makes people comfortable and willing to open up to me; be positive and encouraging; and have fun ... and make the project fun for the client.
- *Environmental Scan*: The government industry is huge, so I have to focus on local governments, as well as nonprofits. Any retiring city or county manager or department head is a potential partner or competitor, and the barriers to becoming a management consultant are minimal. I need to be unique in some identifiable way, based on experience, skills, and personality. I don't have the resources for large projects unless I have reliable partners. I'm not real comfortable in highly political settings (e.g., strong mayor in a very large city). If I don't maintain a work–life balance, I'll fail at my work.
- *Strategies*: Focus on city manager communities under 100,000 population. Focus on three key services (strategic/functional business planning, organizational improvement, and public policy analysis). Continue past competencies in budgeting, revenue enhancement, and cost of service. Keep to the Southeast, Southwest, Midwest, and New England, where I have a history and can easily reach clients via the Atlanta airport. Target my marketing efforts and find ways to let the market come to me. Align and partner with other sole proprietors and firms, like the Mercer Group, but keep the engagement team small (three to five people) so I can consult and not just manage consultants. Be flexible and reinvent myself as needed to serve the market and continue my business.
- *Business Goals*: Keep the business viable for as long as I want to work. Have a great time being a management consultant. Keep all service areas refreshed so the engagement list for each service does not have gaps in work longer than two years. Have every client provide a positive reference due to exceeding their expectations. Maintain a list of responsible and reliable partners and use them periodically so they have a financial interest in my practice. Keep on top of trends and changes in the state and local government market through ongoing research, conferences, publications, and books.
- *Business Objectives*: Win four to six manageable and reasonably-sized projects each year (some alone and some with partners). Bill at least 80 hours per month, while working no more than 160 hours in the heaviest month.

Maintain a fairly even monthly cash flow. Realize 100 percent of my standard billing rate.

- *Personal Objectives*: Be a great husband, father, son, brother, and friend. Never miss church or a family/child event due to the business. Keep running and playing soccer. Take at least four weeks off per year, plus Christmas and Thanksgiving weeks.

Twenty-one years into ECA Strategies, these business plan elements are holding up well. However, achieving eighty billable hours month after month is a tough challenge.

TIPS FOR SOLE PROPRIETORS AND SMALL FIRMS

Sole proprietors and very small firms (e.g., under ten employees) need to carefully define their services and market focus in order to conserve resources in developing client relationships, generating work, and managing and delivering engagements and services. Key objectives are to create a balance and consistency over time, so the firm remains a viable business entity without exhausting the leadership team and staff. Typical strategies for sole proprietors and small firms when developing their niche in the market are listed below:

1. Know yourself first and well before you define a strategic direction for your firm. For the cost of an average lunch, you can take the Gallup Strengths Center's Clifton StrengthsFinder test to identify your five main strengths and see how these strengths match with your role as a professional service provider.[6]
2. Focus on a single industry or on two to three related industries or industry segments (e.g., local governments, utilities, and school boards).
3. Narrow the geographic focus (perhaps to your home city, county, region, or state).
4. Present a small number of core services (three to four) to become an "expert" and to facilitate application of a culture of internal discipline and control. Having several service alternatives also provides options when specific services rise or decline in value to potential clients or because of industry issues or events.
5. Define who will perform practice oversight, marketing, engagement management, engagement staffing, and administrative support roles (a sole proprietor would be like a one-person band in this regard, but with some outside help through contract staff and associated firms and individuals).
6. Differentiate the firm from potential competitors through branding and communications. The Mercer Group, for example, starts off the final slide in its presentation to potential clients with an explanation of "Why We Are Different."
7. Emphasize referrals and targeted marketing, rather than mass marketing approaches.
8. Seek to generate repeat business with existing clients.

9. Develop a broad enough client base to insulate the business if a major client moves on or closes its doors.
10. Develop a small group of mentors to review your strategic plan and evaluate your ongoing business success, then use your mentors to comment on service ideas, business problems, and strategic issues.

WRAP-UP

Developing and maintaining a strategic business plan can be a daunting task, particularly for a sole proprietor or small firm. But it is important to lay out at least a minimal level of strategic planning for the firm and, if it is a larger firm, for each practice area.

Once prepared, the strategic plan should be updated each year based on a retreat or quiet time that disconnects you from the immediacy of the business. The update should include the following:

- A review and assessment of year-end results, for the firm as a whole and by practice areas, and development of corrective action plans if stated goals and objectives have not been achieved.
- Employee interviews or surveys to measure satisfaction with employment, growth and development, compensation, benefits, and the like.
- A review of the vision, mission, and values statements to ensure they still apply as written.
- A mini-environmental scan to ensure you are up to date with market trends and the competition, as well as to identify new business opportunities.
- A review and revision, as needed, of strategies and initiatives.
- An update on progress toward accomplishment of goals and development of new goals, as needed.
- An update on progress toward accomplishing objectives and development of an action plan for those that are behind, need revision, or no longer applicable, as well as development of new objectives, as needed.

Every three to five years, or as dictated by changing circumstances, the sole proprietor or firm should completely redo the strategic plan and, if needed, reinvent the firm to ensure its continued viability in the marketplace. Barry Mundt found that his management consulting practice tended to shift in some dramatic ways (e.g., new services, new assignments, new staff, relocation, new firm name) every three to five years and he wanted to be ahead of the curve to ensure he influenced these changes to his benefit. Steve Egan and Frank Smith have experienced the same change timeline in their management consulting practices. Frank started out working for manufacturing companies and then consulted for manufacturing companies. Because the manufacturing industry in the U.S. shrunk from 40% (in 1982) to 12% (in 2014), Frank redirected his consulting practice to the healthcare industry.

ENDNOTES

1. Moore, Thomas. 2008. *A Life at Work: The Joy of Discovering What You Were Born to Do.* Page xv. New York: Broadway Books.
2. Pfeiffer, William J., Goodstein, Leonard D., and Nolan, Timothy M. 1989. *Shaping Strategic Decisions: Frogs, Dragons, Bees, and Turkey Tails.* Page 12. Glenview, IL: Scott, Foresman and Company.
3. Mercer, James L. 1991. *Strategic Planning for Public Managers.* Page 5. New York: Quorum Books.
4. Pfeiffer, Goodstein, and Nolan. Page 12.
5. Adams, Scott. 1996. *The Dilbert Principle: A Cubicle's-Eye View of Bosses, Meetings, Management Fads, and Other Workplace Afflictions.* Page 162. New York: HarperCollins Publishers, Inc.
6. Rath, Tom. 2007. StrengthsFinder 2.0. New York: Gallop Press.

2 Developing a Culture of Internal Discipline and Control

Rules for Developing a Culture of IDC

Rule 2: Develop and implement a system and culture of internal discipline and control (IDC) to ensure consistency of service, efficiency of operation, and quality and reliability of deliverables. Then, train, mentor, and monitor personnel regarding engagement management policies and procedures.

INTERNAL DISCIPLINE AND CONTROL DEFINED

The authors contend that of the 9 Rules presented in this book, internal discipline and control (IDC), is at the heart of firms that deliver quality results consistently, efficiently, and reliably. These firms establish and then follow their own rules and place a high value on doing so. They have a passion for doing things right and for doing the right things. It's in their DNA. It's their secret sauce.

This chapter addresses the issue of how to achieve sustainability in a firm's delivery of professional services—the first time and every time. Customers judge many firms that deliver professional services as good or good enough. The authors have observed that many firms are good by accident; that is, several factors converge that just happen to produce the desired result. However, such "factors" can't be relied on to produce those results consistently. Others were good once and built substantial good will, only to have successive management teams compromise the existing internal discipline and control culture and fail to leverage the firm's good reputation.

Many professional service delivery firm owners and managers want to know how to instill an IDC culture and its related value system. They want assurances that the firm's professionals will do what they are supposed to do. This chapter defines what IDC is, explains why it is important, and introduces a set of cultural levers that, when deployed, will reinforce behavior that delivers quality results consistently, efficiently, and reliably.

First, let's look at some definitions:

*   *Internal*: Of, relating to, or located within the limits or surface of something; residing in or dependent on the essential nature of something; located, acting, or effective within the body.
*   *Discipline*: Training that is expected to produce a specific character or pattern of behavior; a set of rules or methods; controlled behavior resulting from disciplinary training.
*   *Control*: The power to regulate or guide; a holding back; a restraint, a curb; a means of restraint; a check.

Therefore, internal discipline and control, in the context of professional service delivery, can be described as a *firm's values, expectations*, and *philosophies* related to policies and rules regarding the behaviors to be exhibited and actions to be taken by its personnel before, during, and after the service delivery process. IDC relies on proper personnel training prior to the delivery process, as well as confirmation during and after the delivery process to ensure that the expectations and rules have, in fact, been met.

WHY IS IDC IMPORTANT?

IDC is the key to delivering high-quality service to clients, on time, and within budget—consistently, year after year. Manufacturing firms around the world have had to adopt an IDC philosophy and culture over the past half century, if only to survive. Health care organizations, such as hospitals, have moved toward IDC as a means of improving outcomes and staving off lawsuits. An entity that successfully puts in place an IDC culture and related methodologies likely will be recognized in the marketplace as a reliable and consistent provider and will tend to operate more effectively (do the right things) and efficiently (do things right)—thus improving the financial picture.

IDC—THREE DIMENSIONS

IDC may be considered as having three dimensions in the context of professional service firms. They are represented here as dimensional, because they are not mutually exclusive. That is to say, the development of each needs to be integrated with the others, and not thought of as a separate component. The three dimensions are discussed here in the context of certification or registration, as these are terms that have common understanding in the working world. The three dimensions are industry (or profession), services, and internal processes.

IDC—INDUSTRIES AND PROFESSIONS

The first dimension of discipline and control is a standard or body of knowledge and experience, as it relates to an industry or profession. The following are examples of industries and professions that have certifying organizations:

- Accounting
- Aerospace and defense
- Computer technology
- Business continuity and disaster recovery management
- Environmental professionals
- Government
- Legal
- Medical
- Supply chain and logistics
- Pharmaceuticals
- Trades
- Real estate professional
- Security

Bodies that govern the standards for many of these professions certify practitioners as a benchmark of competency and technical capabilities. Examples of certifying organizations include the American Bar Association, the American Institute of Certified Public Accountants, International Standards Organization (ISO), The Massachusetts Board of Medical Examiners, The State of California Professional Engineers Board, and National Association of Realtors. While these organizations maintain the standards for certification, ultimately it's often an individual who becomes certified. Examples of certified individuals or companies may include the following:

- Certified Public Accountant (CPA)
- Certified Management Consultant by the Institute of Management Consulting (IMC)
- Aerospace Standard 9100 Registered (certified as meeting the 9100 Quality Standard, AS9100)
- Microsoft Certified Information Technology Professional (MCITP)
- Business Continuity Management Systems Implementer (BCMSI)
- Certified Environmental Professional (CEP)
- Certified Manager of Quality/Organizational Excellence (CMQ/OE)
- Admission to the Bar (license to practice law in a given jurisdiction)
- Board-certified medical doctor (MD)
- American Production and Inventory Control Society Certified Supply Chain Professional (APICS CSCP)
- Registered pharmacist (RPh)
- Licensed electrician
- Licensed real estate agent, Certified Residential Specialist (CRS, by National Association of Realtors)
- ITIL Service Management Foundation Certification (Information Technology Infrastructure Library)

In summary, we have presented examples of professions that may be engaged in service delivery followed by examples of certifying organizations, and end with examples of certifications that some of these organizations may confer upon the professionals who practice in the various professions.

IDC—Services

A second dimension of discipline and control is one that is service specific. Oftentimes companies establish certifications related to the installation and use of their product(s) to maintain a high level of product quality. Certain software companies, for example, require installers and resellers of its product to be certified to ensure that a high-quality standard is maintained and to validate warranty coverage.

Similarly, some services are certified or specified by industry or professional regulatory bodies. For example, the Financial Accounting Standards Board (FASB) or the Government Accounting Standards Board (GASB) sets down the process and content for financial audits. The U.S. Government Accountability Office publishes *Government Auditing Standards* (a.k.a. the "Yellow Book") that includes standards for performance audits, which are akin to management consulting "studies."

IDC—Internal Processes

The third dimension related to IDC is internal governance—a process to govern how personnel go about delivering the service for which the client is willing to pay. This last area is where the project or program manager is involved directly. The discussion around service delivery process is at the heart of maintaining internal discipline and control. It should be noted that the discipline of project management and its related certification by the Project Management Institute (PMI) was not included in the discussion of IDC industries earlier in this chapter, as it is integral to managing the implementation of processes and the related infrastructure associated with sustainability of process improvements.

The notion of discipline referred to here is not a knowledge discipline specific to any industry; rather, it is a discipline of process and how things get done. The aim of this third dimension is to standardize processes (how things are done) as much as possible, and then ensure that the processes are followed (control). Maintaining IDC means that the outcome of processes performed by all personnel will be highly predictable. Higher predictability means that there will be less variability in the outcome of a process (i.e., the service delivered). Minimizing variability in the outcome of processes means that you can charge a fixed price, because you have a high level of confidence in the amount of time it will take to perform the process.

The notion of applying the principles of project management to professional service firms relates to the need to standardize internal business processes as a way to improve efficiency of the operations. This increased attention on improving efficiency is in direct response to customer demand for firms to be able to deliver a reliable quote for services in response to a request for proposal. Recall from the Preface that *efficiency* plus *effectiveness* yields *productivity*. Customers are no longer accepting responses from professional service firms that "it takes as long as it takes, and therefore we can't provide a quote for services other than our hourly rate for time and materials." This is especially true in the public sector, where a winning proposal is one that has both the lowest cost (efficient) *and* is responsive (effective).

The new paradigm requires service delivery companies to identify and define, with sufficient detail, all of the processes in which it engages to deliver its services effectively. This means that a service delivery organization must standardize as many of its internal processes as it can. It then must require its associates to adhere to the business process standards whenever they perform the work associated with any of the business processes.

The notions of (1) customized service and (2) standardized processes seem to be in direct conflict with each other. Yet, those firms that are able to achieve a dynamic balance between defining and standardizing their processes and adhering to the standards (thus managing costs), *without sacrificing the quality of service inherent in customized delivery*, will be the winners in their industry or service segment. To be sure, this notion is exactly the discipline that came of age in the manufacturing industry some years ago. It is now being extended into professional service delivery.

The astute reader will recognize immediately that what is being discussed here represents a significant change in the culture of how things are done in their respective profession, particularly for sole practitioners and small firms. Relatedly, they will anticipate correctly that their peers and subordinates will resist embracing this new paradigm—with intensity. To be sure, leverage will have to be applied to exact a change in behavior.

ORGANIZATIONAL CULTURE AND RESISTANCE TO CHANGE

Let's define what we mean by the culture of an organization. *Culture* is defined as how one goes about getting things done in his or her workplace. It is also defined by the nature and existence of relationships and management philosophy. Guido Slangan, professor of organizational management at the Hartford Graduate Center in Hartford, Connecticut, referred to the "Emergent System."[1]

Every company has the formal organization chart, with which most readers are familiar, and also the informal organization structure that is not reflected in the formal organization chart. The emergent structure may be defined by family, common alma mater, friendships that have evolved since childhood, ethnic background, or some other common element.

Physical structure may also play a role in defining the culture of an organization. Some companies have migrated from having offices and cubicles assigned to specific individuals, to spaces assigned to whomever happens to be working in the office that day. For example, some companies have determined that it is not cost effective to pay for space that has low utilization because the person who occupies that space is in the field working at a client site for many of his or her yearly hours. In another case, the presence of an air hockey or foosball game in the middle of a work area may define the company's culture.

Many companies have documented policies and procedures, but the culture may be one where the policies and procedures are not enforced. In fact, the existing documentation may be viewed as nothing more than bureaucratic red tape that gets in the way of getting a job done. Other cultures may have documented policies and procedures because they are required to have them to be in compliance with some governing authority, but give lip service to the notion of actually enforcing them unless forced to do so. An organization whose culture is defined as having good IDC is one where internal policies and procedures exist, are maintained and updated, have a procedure for doing so, and all of the policies and procedures are adhered to consistently; the policies and procedures describe accurately how things get done.

It is the natural order of things for people in organizations to come up with their own way of doing things if the preferred way is not defined, communicated, and enforced. This is most acutely apparent in large organizations with multiple sites or organizations that have grown largely through acquisition. Multiple sites where employees don't have contact with each other do not have the benefit of learning from each site's best practices. In fact, some companies may have reward systems that discourage site managers from sharing best practices with each other. Our experience is that the more heads that are in the game, the better the outcome. Consequently, we would argue that site managers should not only share best practices, but also the sharing behavior should be encouraged and rewarded. The need to employ different kinds of leverage begins to become apparent when viewed from this perspective.

Of course it is not reasonable to ask employees to change the way they do things, unless we first define specifically the behavior we want them to exhibit—how we want work to be done. Further, our experience is that many organizations have neither well-defined processes for how work is to be done, nor clear documentation explaining the work processes. Discipline can't be instilled and enforced if we don't make clear what is expected or how we expect tasks to be performed. Further, a common standard must exist. It should explain exactly how the work is to be performed and be clearly understood. Lastly, practitioners must also understand how they will be measured against the standard and how their score will be calculated.

ESTABLISHING AN IDC PHILOSOPHY AND CULTURE

An IDC philosophy must be established at the top of an organization, communicated and reinforced effectively, and demonstrated and modeled (indicated by behavioral example) regularly to all employees. This philosophy is first expressed through the firm's expectations, policies, and related rules regarding how engagements are to be conducted, managed, and controlled—thus reinforcing the desired ways of conducting engagement delivery. The power of setting and articulating expectations needs to be emphasized here by recalling the results of an experiment known to most educators: the Pygmalion effect.[2] Robert Rosenthal and Lenore Jacobson published the results of a study in 1968, which showed that people rise to *your* level of expectation. Many sociologists postulate that people internalize the expectation or label assigned to them and behave accordingly.

First, we have to define how things are going to be done through the establishment of policies and procedures. The procedures need to reflect what are sometimes referred to as best practices: how things are going to be done and the behaviors that we want our professionals to exhibit. Then we have to employ cultural levers to reinforce the desired behaviors. But before we can begin the process of defining and documenting best practices, we have to agree on a framework for managing the documentation of processes, including management of the process for changing a particular policy or procedure and communicating the change to relevant personnel in the organization.

DOCUMENTING AND COMMUNICATING THE IDC METHODOLOGY

Internal discipline and control needs to be documented in writing and communicated to an entire organization in order to be effective. One of the most effective ways to communicate IDC expectations is to document "the way things are done around here" in the form of policies and procedures. The terms *policy* and *procedure* are often used interchangeably or together as inseparable concepts, leaving the reader to infer that the two words together are one (much like the terms *sales* and *marketing*). But they are not the same thing.

POLICY DEFINED

The World English Dictionary defines *policy* as a "plan of action adopted or pursued by an individual, government, party, business, etc." In the context of a professional service firm, a policy is a document that provides guidelines about a firm-sponsored system or practice. In essence, a policy statement reflects the culture your firm wants to adopt.

Policies are specific in nature and content. They are "published" as a means of communication to employees of the firm. Whenever possible, statements of policy should have consistency in format (see Figure 2.1 for a suggested policy format). A policy statement should contain at least the following sections:

1. Purpose—The purpose of the policy, stated in a single sentence
2. Policy Statement—A statement of the policy that is desired or mandated, in whatever length is appropriate to communicate its intent
3. Scope—The scope of coverage, which may refer to specific entities (e.g., departments) within the firm or specific practice area(s) of the firm
4. Responsibility—The title of the position(s) that hold primary responsibility for ensuring that the policy is implemented
5. Cross-Reference—Identification of the procedures (if any) that are designed to ensure implementation of the policy

PROCEDURE DEFINED

Procedures depict the steps one should follow in carrying out a guideline specified in a policy statement. Like policy statements, procedures are "published" and, whenever possible, have consistent format (see Figure 2.2 for a suggested format). Procedures are specific in nature and content. They should have at least the following sections:

POLICY		Number
		Revision Date
		Page 1 of X
Subject/Title		Approved by:
1. Purpose:		
2. Policy Statement:		
3. Scope:		
4. Responsibility:		
5. Cross-Reference:		

FIGURE 2.1   Sample format of a policy statement.

PROCEDURE		Number
		Revision Date
		Page 1 of X
Subject/Title		Approved by:
1. Policy Reference:		
2. Purpose:		
3. Procedure:		
4. Scope:		
5. Responsibility:		
6. Applicable Documents:		

FIGURE 2.2   Sample format of a procedure.

1. Policy Reference—Refer to any policy statements that apply to this procedure (e.g., policy number, latest revision date, and subject/title)
2. Purpose—The purpose of the procedure, stated in a single sentence
3. Procedure—A statement of the desired or mandated procedure, in whatever length is appropriate to communicate its steps (typically displayed in two columns, the first column identifying the position responsible for performing the step and the second column specifying the step to be performed)

4. Scope—The scope of coverage, which may refer to specific entities (e.g., departments) within the firm or specific practice area(s) of the firm
5. Responsibility—The title of the position(s) that hold primary responsibility for ensuring that the procedure is implemented appropriately and properly
6. Applicable Documents—Cite any supporting documentation for this procedure

POLICY AND PROCEDURE DOCUMENTATION

The authors have heard comments over the years that documentation of policies and procedures is cumbersome, difficult to maintain, and never gets used. We would argue that a professional service delivery firm that does *not* document its internal policies and procedures (i.e., its rules of governance) is not sustainable. Indeed, the business press is filled with stories over the years of companies that have ceased operations because they either did not have proper policies and procedures in place or didn't adhere to the ones that were in place. The governing structure becomes especially important when a professional has to adapt the engagement plan to the real world, which is the focus of Chapter 5.

A group of professionals may discuss and agree that they want things done a certain way. Memories are short, however, and not everyone may remember what was agreed to during the discussion "way back when." Adherence to discipline begins with writing down exactly how the team of professionals is going to deliver its professional service. Our recommendation is that the very first document that is created should be a statement of purpose. We call this the "genesis" document. In it, the firm's management agrees that they should establish and maintain a documented set of policies and procedures that govern how they are going to perform functions within the scope of their professional service delivery.

The purpose of the genesis document may be stated as, "This policy provides for a system of management documentation that serves to supply guidelines to firm personnel in the conduct of the firm's business." The purpose section of the genesis document should be followed by definitions of both *policy* and a *procedure*.

The next document should spell out exactly how the firm will go about changing any part of policy and procedure documentation, if it becomes necessary (this is sometimes referred to as *change control*). The policy should also spell out the frequency of review and updating. Other policies, for example, may detail the following:

- Content and use of marketing materials in outreach campaigns
- Development and use of standard firm documents (e.g., letters, proposals, deliverables or elements of deliverables)
- Standard billing rates for all staff
- Situations where discounted or premium rates are appropriate
- Process of proposal writing and review prior to submission
- Use of subcontractors on engagements

Finally, it should be noted that the firm's policies and procedures should include not only a statement of the types of client engagements that are deemed acceptable, but particularly those that are *not* to be pursued and accepted (such as unethical, high risk, and/or conflict of interest situations). That procedure document should outline the specific steps to follow in determining if a potential client is someone with whom the firm wants to do business.

DEFINING BEST PRACTICES

One of the ways that we have been successful in getting people to adopt how things will be done is to seek out best practices and make them part of the IDC process. We do so by identifying subject matter experts (SMEs) within the ranks of employees. These people are recognized as having extensive experience working in the area to be documented—personnel who have tweaked a process an untold number of times over the years, until the process is fine-tuned. We have found that including SMEs in the process of defining a best practice ensures a higher probability that the best practice will be embraced, because the people who do the work or are responsible for getting the work done have defined how it will get done. One process for defining and documenting best practices is set out in Table 2.1.

Sometimes the process of achieving consensus in defining a best practice is messy. Some participants may come to the table with preconceived notions, hidden objectives, and/or residual emotional baggage. A skilled facilitator will be able to get the parties to resolve differences. Sometimes, the process highlights an individual who will simply not fit in with the new culture, despite the best efforts of all involved. It is usually in the best interests of the management to identify individuals who don't or won't fit in and determine their future with the organization as soon as possible.

Once the firm has defined the processes that govern how business is going to be conducted, including standards for documentation of the same, it needs to employ a mechanism to ensure that its professionals learn, use, and adhere to those standards. We have enjoyed success in this regard through the deployment of *cultural levers*. A Manual of Business Processes documents how business is going to be conducted. In fact, publication of the manual represents deployment of one cultural lever. Additional levers need to be deployed to increase the probability that a culture of IDC takes root and is sustained.

EMPLOYING CULTURAL LEVERS

The concept of *cultural levers* was first introduced to one of the authors by Richard Kristensen, managing partner of Harris Chapman,[3] a management consulting firm. Several key ingredients are related to the deployment of cultural levers:

1. Deploying only one lever or a couple at a time will not produce sustainable results.
2. Select multiple levers that are applicable to the goals you are trying to attain.
3. The impact of deploying multiple levers is not linear; it is exponential.

Eight cultural levers are discussed in this section: customs and norms, objectives and measurement, ceremonies and events, leadership practices and behaviors,

TABLE 2.1
Steps in Best Practice Definition and Documentation

1. Set goals and/or objectives for the business process(es) that you intend to document. An example might be to define and document the best practice for creating bills and sending them to customers according to a set schedule, and completing this definition and documentation exercise in four hours.
2. Define the scope of the process that needs to be documented. Management needs to determine the elements to be included in the scoping exercise. For example: *Start with the point at which a client makes first contact with the firm and end when the client exits.*
3. Divide the process into components, once the scope is agreed to. For example: Individual components of the business process might include Billing, Dunning, Report Preparation, etc.
4. Identify SMEs for each component of the process; then form a team of people for each component.
5. Schedule separate work sessions for each of the process components; the length of each work session will vary depending on the number of functions you are trying to standardize and the amount of variability between each location performing the function. The amount of effort required to define and document the component of a process where two departments are being merged is very different from standardizing the work of fifty sites across the United States.
6. Have a disinterested third party facilitate the work session, with the clear goal of defining the best practice for performing that component of the process.
7. Upon completion, have the team conduct a report-out session to members of management, including documentation of how the business process will be conducted.
8. When all of the teams have completed their respective sessions, each team's documentation should be assembled into a Manual of Business Processes. The individual firm may decide to use this manual to be the de facto set of procedures or it may include it in the procedure documentation by reference. Regardless of how it is incorporated, the manual's contents will be the "bible" for how the business processes will be conducted in the future. Small firms or sole proprietors may need to assemble a scaled-down version of a manual.

rewards and recognition, communications, physical work environment, and skills and knowledge training. An explanation of each follows.

Customs and Norms

A custom is a habitual practice, or the usual way of acting in given circumstances. Customs define "how things are done around here." Customs are usually not documented and may even be unstated. An example of a custom is the use of certain slang expressions in the U.S. Navy. Specifically, the phrase "very well" is an expression of acknowledgment that an officer may say to a subordinate or enlisted person. It is not something an enlisted person would say and would be frowned upon by officers if he or she did say it in the presence of an officer. This "rule" doesn't appear in the navy's *Bluejackets Manual*, the basic handbook for U.S. Navy personnel. It is learned in the field through custom and repetition.

A *norm* is an established standard of behavior that is shared by members of a group and to which each member is expected to conform. Norms include such things as timeliness, accuracy, compliance, attention to detail, and customer orientation. Norms are likely to be documented—if not in the organization's policies and procedures, then in a memorandum to the organization's personnel.

The goal is to learn standards of behavior through intentional means and not by accident, such that undesired behaviors or the old way of doing things is eliminated. The application of this lever rewards professionals who exhibit the new desired behaviors.

We have mentioned earlier that sometimes a firm's success is accidental. Sometimes that success becomes the justification for and dictates how things are going to be done in the future. This is a risky proposition if the reason for the success is accidental, as one may have low confidence that the approach will be successful again or that it is the correct one for all future engagements. Furthermore, if an analysis isn't performed to identify the factors that contributed to the firm's success, it will be difficult to replicate the success with consistency.

If we want to create a specific culture in an enterprise, we first have to define what that culture looks like, as well as the specific kinds of behavior that we want employees to exhibit. To say, "I'll know it when I see it," is an accidental approach that represents a reactive mindset. A reactive mindset does not contribute to results that are sustainable. A reactive mindset will only produce accidental results, regardless whether they are good or bad.

The earlier discussion of the need for documenting policies and procedures, as well as how to engage personnel in the process of defining how processes will be performed, will help to produce the specific behaviors that you want your firm's professionals to exhibit. In addition to documenting policies and procedures, your firm should also provide a vehicle for personnel to propose changes to existing policies and procedures. This will enable the organization to improve business processes as conditions or experience dictates. It needs to be appreciated that a best practice adopted on the first day may be improved upon over time.

Many companies have established procedures where changes to an existing process may be submitted from the field. For example, a Change Control Board may be empowered to review proposals to change business processes. The Change Control Board notifies the change request originator of its decision and the timing of when the change will go into effect, if approved. A change transmittal notifies recipients that changes have been made to existing policies and procedures, including chapter, section, page number, and paragraph, and that specific wording has been changed from "xxx" to "yyy." Finally, the person responsible for performing this task needs to know it is their job and management needs to ensure that it is performed.

Objectives and Measurement

Objectives need to be developed that are specific, in that they describe the behavior you want your professionals to exhibit. They also need to be associated with each operation you are trying to change. The objectives and the related measurement of performance will help to reinforce the changes that are desired. For example, firms that establish an internal audit function that checks adherence to internal policies and procedures, coupled with a compensation system where the bonus payment is tied directly to the achievement of specific audit scores, ensure a higher probability of adherence to the expected standards. Our experience is that compliance is achieved when the performance metrics and the precise calculations to be used are clearly defined and unambiguous. In a small firm, the president or a partner might assume the role of internal auditor.

Ceremonies and Events

Ceremonies and events include special assemblies where awards are given out. These include recognition events for teams and individuals who demonstrate behaviorally the organization's values. A firm should establish events that celebrate and reinforce new desired ways of doing things. For example, we have seen professional service firms where the CEO or a partner makes it a point to visit key sites to personally praise an individual who has excelled in performing a designated process; this tends to be done amid much pomp and fanfare, in the company of the individual's peers. It is further reinforced with the award of additional cash on the spot (sometimes called a Spot Award). Such impromptu events may be followed with the publication of an announcement, together with pictures in the company newsletter; this reinforces the message that the company is serious about the achievement of standards of excellence and is willing to put its money where its mouth is.

We heard of one company owner who was struggling with the viability of the company as a going concern and feared it would fail soon. A shop-floor worker walked into his office with an idea that the owner felt would save the company. The owner wanted to reward the worker for the idea immediately but could only find a banana in his desk drawer. That act resulted in a Golden Banana award for any ideas that benefit the company measurably.

Leadership Practices and Behaviors

Demonstrating behaviors of the organization's success model is a key lever. Company leaders must show what is really important and how to get ahead by identifying and eliminating self-defeating management practices. This involves the replacement of outmoded practices with conscious efforts to model new desirable behaviors, by rewarding those who demonstrate the right behaviors, and penalizing those who do not change undesired behaviors.

Our experience has shown that rewarding *progress improvement* yields significant performance improvement, versus rewarding only "perfection." It is not enough to simply define best practices and then implement a performance measurement system, give rewards, and assume that behavior will change. Personnel need to be coached toward the desired behavior. More specifically, acknowledging and praising progress made toward a specific change in behavior is a very powerful motivator.

Coaching models exist in various forms. We recommend a simplified, multistep approach where the coach defines in clear detail the behavior that is expected and the behavior that is observed. The coach engages the subject to discuss the expected versus observed behavior, and then asks the subject to explain the rationale for the observed behavior. The rationale will help to illuminate the underlying values that drove the observed behavior. The desired goal is to gain alignment between the coach and the subject so that each shares the same values. Once alignment of values is achieved, trust is established and the relationship between the coach and the subject is strengthened. The two can then proceed to collaborate in mapping out a plan of action to change the observed behavior so that it meets expectation. Having the subject participate in developing the plan elevates the probability that they will take ownership of the plan, commit to getting it done, and be accountable for the results.

Finally, using role playing will help identify areas where the subject needs to show improvement and where the coach needs to focus to develop skills further. This exercise will also help to determine a subject's readiness to assume certain roles in the engagement delivery process.

Rewards and Recognition

Rewards may be both financial and nonfinancial. Sometimes rewards can be in the form of a certificate recognizing an achievement. Recognition may be more effective if it is specific to the change goals and objectives that have been set. To be sure, care should be taken to eliminate rewards and recognition for methods and procedures that are no longer desired, and replaced with a rewards structure that reinforces the new desired ways of doing things.

Communications

You can't overdo communications that provide information needed to perform job tasks. These communications should be in a form that allows all persons to feel involved, informed, and in the know about what is going on in their organization. Delivering the communication in new ways demonstrates the firm's commitment to change. Using multiple channels allows the organization to be consistent in its messages before, during, and after the change. Lastly, the communication needs to be two-way, such that feedback is solicited regarding the clarity, credibility, and impact of each message.

Physical Work Environment

Arrangement of the workspace and physical organization needs to be considered. We must ensure that people who need to work together have accessibility to each other in a way that is convenient and consistent. Rearranging workspace and/or relocating personnel to reinforce desired changes in operating practices may have a significant impact on performance. For example, we have implemented a new management infrastructure at client companies where office personnel have suggested changes in the physical work environment. In one case, workers recommended rotating their desks and chairs a mere 90 degrees. This simple change delivered substantial productivity improvements. Of course, sometimes changes in process flows have required the removal of walls or repositioning workers closer to the recipients of their end product.

In another example, the authors worked at a firm where professionals of three different functional areas (audit, tax, and management consulting) were housed in separate floors of a local office building. A decision was made to rearrange the seating assignments, such that professionals were co-located according to the industries they served, rather than the functions they performed. The principle of propinquity holds that people tend to develop relationships with others who are physically closest to them. In the old model, all the tax professionals ate lunch together, all the auditors ate lunch together, and all the management consultants ate lunch together. In the new model, lunch gatherings began to change, such that all three functional areas that served a particular industry began to eat lunch together. Conversations tended to focus on specific clients and how each of the respective professionals could contribute to delivering the best service. The office realized a significant increase in revenue.

Skills and Knowledge Training

Skills and knowledge training involves preparing people for successful performance in current tasks, as well as for assuming greater responsibilities. It requires replacement of general education and one-size-fits-all training with "just-in-time/just enough" skill and knowledge development activities. The trick is to provide real-time, hands-on experience with new processes and procedures.

For example, an organization may designate personnel in the field as SMEs for particular components of a business process. A SME may be dispatched to a particular site to help remediate the failure of a business unit to achieve minimum best practice standards. Requiring specific individuals to pass a mandatory exam covering best practices, conditional on a bonus being paid, can reinforce adherence further.

Training and development should include not only the personal and technical skills needed to perform appropriately on an engagement, but the IDC expectations, policies, and procedures, as well. Each service delivery organization needs to institutionalize its complete set of best practices, including policies and procedures, once it has defined them. Pay, recognition, and promotion systems typically are designed to reward consistent excellence in engagement delivery, but also they should include adherence to the IDC policies, procedures, and standards. Engagement performance reporting and management practices should include an assessment of how well an engagement team and its individual members have abided by the IDC policies, procedures, and standards *and* met the expectations of the firm's leaders. There is an added benefit to your firm's professionals being knowledgeable in all aspects of the firm's policies and procedures. They will be able to coalesce as a productive team much more quickly than a group that is without the common base of understanding. This *common base of understanding* will be a key differentiator when your clients assess whether or not you have met their expectations.

A well-designed personnel assessment system is one that has improvement of personnel behavior as a core value—as opposed to catching personnel doing something wrong and punishing them. Coupling the identification of deficiencies with mitigation efforts and then rewarding improvement through a coaching model usually results in an increase in the desired behavior. This approach has a much higher probability of changing (and improving) behavior.

We have experienced success utilizing a Skills Flexibility Matrix (SFM). An SFM shown in Figure 2.3 is simply a two-dimensional spreadsheet showing two dimensions. The names of the people being considered for a particular position are listed along the left side, and all of the skills needing to be resident within the scope of an organizational unit are listed across the top. Each person is evaluated regarding the level of mastery for each skill using a numerical value. A zero means the skill is not applicable to the individual, 1 means needs training, 2 means entry level, 3 means advanced knowledge, and 4 means mastery and able to instruct others. Each cell in the spreadsheet is programmed such that entry of a number 0 through 4 will also change the cell color to clear, red, yellow, blue, and green, respectively. A completed SFM becomes a visual aid for the organization's manager, in that it allows him or her to see at a glance where skill deficiencies exist within their area. It will also allow the manager to begin tailoring a training program that focuses on the skill gaps that exist. The overall functioning of the unit's operation is lifted when personnel complete the training program. The SFM will allow the manager to focus on improving the overall functioning of their area.

CULTURAL CHANGE

Many organizations ignore or shrink from cultural change, because the subject is viewed largely as an emotional concept that appears ephemeral and difficult to manage. In fact, culture is made up of "tangible" elements: history, organizational structure, people, operations, equipment, systems, and procedures. It should be noted that an organization committed to improving these elements must manage, continuously, as many of the cultural levers as possible. Change that is not managed continuously will not yield sustained results.

ORGANIZING AND STAFFING FOR IDC

Human resource management, in particular the recruiting and selection process, can be used to instill an IDC culture in an organization.[4] How work is organized managerially needs to include the flow of authority, responsibility, and accountability. The organization should be structured such that it will reinforce desired operational and cultural changes. Examples of such a structure include the following:

- Establishment of client service teams
- Eliminating excessive layers of management
- Centralizing or decentralizing work, as necessary
- Changing the role of the supervisor/manager to coach personnel, rather than to boss them

We list the recruiting and selection process first because that's how to assemble a cadre of personnel who have the potential to accept and work successfully in the

Skills Flexibility Matrix

Flexibility Rating:		4 Proficient	Trainer for Future Managers in specific skill set
		3. Average	Strong in specific skill set, no guidance needed
Department/Area:		2. Aware	Needs work in specific skill set
		1. Unskilled	Needs to be trained
Managers:		NR	Not Rated

Date Completed:

Employees	Months Eployed	Position	Skill 1	Skill 2	Skill 3	Skill 4	Skill 5	Skill 6	Skill 7	Skill 8	Skill 9	Skill 10	Skill 11	Skill 12	Skill 13	Skill 14	Skill 15	Employee total	Employee Average
Employee A	17	Position A	2	3	3	3	3	3	3	3	3	3	3	3	3	3	1	42	2.8
Employee B	84	Position B	2	3	3	3	3	3	3	1	3	3	3	3	3	3	1	40	2.7
Employee C	0.5	Position C	1	2	2	1	2	1	1	1	1	1	1	2	2	1	1	20	1.3
Employee D	0.5	Position D	3	3	3	1	3	3	3	3	3	3	3	3	3	1	1	39	2.6
Activity Average			2.0	2.8	2.8	2.0	2.8	2.5	2.5	2.0	2.5	2.5	2.5	2.8	2.8	2.0	1.0	59.0	2.4

FIGURE 2.3 Skills Flexibility Matrix.

desired IDC culture. The recruiting and selection process is not limited to employees but extends to contractors and subcontractors as well.

Use of the SFM is also an opportunity to determine if your firm's values and those of the candidate are aligned. Uncovering those "soft" aspects of a candidate's makeup can be a challenge. We turn once again to Richard Kristensen, who eschews the typical approach to questioning such as, "Tell me what your strengths and weaknesses are." Instead, he advocates starting with open-ended questions that can't be answered yes or no. His approach is to not stop at the first answer. Instead, he says you should continue to probe for more and more detail. The purpose of this approach is to determine the candidate's depth of experience and understanding. It provides insight into their proclivity to appreciate subtle nuance; alternatively, it will surface the lack of it. A wise friend once observed, "Fabricators have no details."

SETTING UP THE SUPPORTING TOOLS AND TECHNOLOGY

Service delivery professionals need a set of tools and technologies that will enable them to perform their work. Providing tools and technology to service professionals allows them to collaborate on engagements, thus ensuring consistency and efficient service delivery. It also provides managers the ability to track and measure adherence to established policies, procedures, and standards.

TOOLS AND TECHNOLOGY

A number of alternatives are available for use in managing professional service delivery. Office technology is constantly evolving, and much of that evolution is aligned with a teamwork model that stresses collaboration. Applications that allow for real-time communication between team members working on an engagement are preferred. Further, products that track the number of hours associated with a single project, as well as the availability of hours by skill classification in the resource pool of the firm, are also preferred. Examples include the following:

- Smartphones
- Email
- Scanning hard copy documents and other artifacts
- Electronic calendars that allow team members to view each other's calendar
- Electronic white boards
- Digital survey instruments
- Storage of digital artifacts (documents, video, sound recordings, etc.)
- Version control (for computer files)
- Videoconferencing
- Conference calling
- Mobile communications
- Desktop messaging
- Word processing, spreadsheet, presentation, and business process flow documentation to create digital process flow diagrams

Use of the brown paper method is a low-technology tool that enables team members to collaborate real time, while documenting a business process flow. The butcher-type paper comes in a large roll and is about four feet wide. The team members draw a process flow diagram on the brown paper and attach samples of artifacts on the paper using a glue stick. The paper is hung up on a wall so that it is visible to the team. It allows team members to accurately depict the as-is process. It also allows the members to see an entire process from end to end, instead of the limited view of their particular piece of the process. This tool has been shown to be very effective in identifying gaps, duplications, or inefficiencies in an existing process.

TEMPLATES AND MODEL FORMATS

Templates and model formats are handy devices to have in your toolkit. They help to standardize processes and save team members from having to re-create many of the artifacts used in service delivery. Examples of templates or model formats that are commonly used follow:

- Information request
- Meeting invitation
- Meeting agenda
- Meeting minutes
- Status reports
- Letters and lists
- Proposal
- Letter report
- Technical report
- Report/deliverable transmittal letter

Other templates or formats should be identified that will make life easier for your professional service staff.

WALK THE TALK (OR PRACTICE WHAT YOU PREACH)

It is extremely important that all executives and other levels of management abide by the IDC policies and procedures in a professional service firm. If *they* don't walk the talk, then the professional service staff won't abide by the rules either. Thus, executives and management must demonstrate that they are serious about monitoring and enforcing the established policies and procedures. But they also must practice what they preach; if they do, the staff members will be more likely to follow their lead.

TIPS FOR SOLE PROPRIETORS AND SMALL FIRMS

Sole proprietors and small firms have unique challenges in developing a system of internal discipline and control, primarily because they lack the staff and personal time to devote to internal administrative projects like IDC. Some suggestions follow on how to develop a workable IDC for your small firm:

1. Standardize your services so they are easy to explain to your staff and can be repeated from one client to the next. Although standardized to your firm, make them unique in some way so that they may be differentiated from your competitors. For example, Steve Egan's management studies are based on his 50 Management Issues for Organizational Improvement, which basically is a checklist of what is reviewed in a management consulting engagement. Clients seem to respond to a tool set and approach that is thought out well.

2. We strongly recommend that you develop some form of IDC to hold your practice together. Start simply at first with key needs; then expand the IDC as you learn from client engagements. Before you start any client work, develop a few memos or adapt pages from our book for at least these few key topics, such as leadership roles, proposal development and approval, deliverable preparation and reviews, client reporting, and budget planning and reporting.

3. The greatest source of resistance to change might be *you* (and your partners if you have them). Fight against personal and firm inertia. Don't come to rest ... ever!

4. Create customs and norms for your firm or practice that suit the way you think and work. Hire partners and staff who "fit" your style. Being on the same page is more important than being right, perfect, or best.

5. Create a "nest" in a place or an office where you can focus on your work undisturbed. The dining room or kitchen table won't work. Making a client call from the kitchen phone with a couple of screaming kids and a barking dog makes a poor impression and won't get you hired. Work in a professional office space even if you are not residing in Class A office space in New York City.

6. Reward yourself (if a sole proprietor) or your staff (if a firm) regularly, even if it's a small thing like lunch, a $25 gift card, a new briefcase, or a half-day off.

7. Take time to learn and grow (e.g., attend webinars, read books and articles, engage in networking) so you are kept up to date with industry knowledge and become more experienced every day.

8. Discover effective and reliable external associates (e.g., subcontractors, associated firms) so you have support on elements of an engagement that are in areas where you lack strengths. Use them regularly so they have a stake in your practice and your success, and vice versa.

9. Invest, and don't skimp, on key technology tools, such as your smartphone, computer, and printer. Define what you absolutely need to do a great job; then find a way to acquire it. Take a pass on fancy wants and likes.

ENDNOTES

1. Slangen, Guido H. 1985. Class Lecture on *Organizational Behavior*. Hartford, CT: Rennselaer at Hartford.
2. Rosenthal, R., and Jacobson, L. 1968. *Pygmalion in the Classroom*. New York: Holt, Rinehart & Winston.
3. Kristensen, Richard. 2012. *Change Management: Organizational Renewal and Cultural Levers*. Jupiter, FL: Harris Chapman Presentation Document.
4. See: *HR Impact on Corporate Culture*, www.hr.com, July 1, 2005.

3 Selling the Engagement

Rules for Selling the Engagement

Rule 5: Prepare complete and definitive proposals, contracts, and engagement work plans that evaluate and accommodate engagement risks for both the provider and the client, so the client knows what can be expected in terms of scope, work plan, schedule, deliverables, and cost.

DEFINING SERVICES TO BE DELIVERED

This chapter focuses on a number of aspects related to selling professional services. The concept of internal discipline and control (IDC) presented in Chapter 2 extends to the process of selling services as well. We have learned that the more disciplined we are in defining the services we are going to deliver, the better we will be at developing the service, representing it to the client, and delivering results that meet or exceed their expectation. A crisp and clear presentation also enables one to distinguish their firm from the competition. This chapter presents a six-step process for selling engagements. The chapter begins by presenting a process for defining the services to be delivered. This is followed with ideas for identifying and contacting potential clients, responding to a request for proposal, closing the sale, entering into a contract, and what to do if your proposal isn't accepted.

Most professional service firms focus on a particular industry or a function that is relevant across several industries. Successful professional service delivery firms define the services they deliver with precision. That is, the scope of what is and is not included is very well defined and strictly adhered to, unless there is a good reason to modify the scope.

Most successful firms study the market to determine the needs of the marketplace and align its core competencies with those needs. Aligning core competencies means that the professional service firm should translate its core competencies into services.

The more discipline and structure built into the identification and development of services, the greater the likelihood of success. Accordingly, we recommend that firms develop a *service development process* (SDP). Steps included in an SDP may vary according to the discipline being considered. Regardless, we believe that the following steps should be considered at a minimum:

1. Presentation of an idea or design for a service
2. Development and validation of service concept design
3. Creation and testing of a prototype version
4. Testing of a pilot version with a live client
5. Packaging the service
6. Planning and implementing the market launch or rollout
7. Measuring service performance
8. Planning service growth and continuous improvement

These steps are described in the following paragraphs. The reader will be wise to recognize that the definition and development of specific services needs to be done in the context of the strategic business planning process presented in Chapter 1. The specific service ideas that are aligned with the firm's strategy will have a higher probability of being the tactical enablers of the firm's overall strategy.

PRESENTATION OF AN IDEA OR DESIGN FOR A SERVICE

The service development team must begin with an idea or a design of a service to be delivered. Ideas may originate from numerous sources: employees, existing clients, brainstorming sessions, target research, or simply a hunch. Alternatively, sometimes an originator of an idea may go so far as to develop a design for the service in order to give the approving body more information or a better feel for the idea being considered. Regardless of their source, firms should have a well-defined entry point for submitting an idea or a design for a service.

Some firms engage in a hit-or-miss approach to service identification. They don't conduct market research or use facts to support the selection and launch of a service. Some may reason that because a particular service was successful for one client, then all clients in the industry will want it as well. Sometimes they are successful, but often they are not. Some entrepreneurs seem to be able to just stick with it—or maybe have unlimited resources. It has been said that Colonel Sanders launched twenty-one unsuccessful business ventures before launching Kentucky Fried Chicken (KFC). We suspect that most firms do not fall into this category.

We prefer to engage in a process of gathering data and facts gleaned from structured research that confirms the validity of an idea or hunch. The idea should pass a minimum threshold of rigor to determine if the firm is willing to commit the resources necessary to pursue the idea further. It also helps to know what the threshold attributes are (such as alignment with the firm's strategic direction, mentioned earlier). A structured approach will increase the probability that the service ideas have been subjected to an appropriate level of research rigor relative to the threshold attributes. That will result in a higher probability that the service ideas will be selected.

DEVELOPMENT AND VALIDATION OF A SERVICE CONCEPT DESIGN

The idea should be developed into a concept design once the service passes the threshold mentioned above. The concept design should explain what the service is, how the idea can be fashioned into a service, how it can be delivered, how it can deliver value in the form of results, and why clients will be willing to pay for the service. Some firms set a minimum threshold that defines the rate of return that clients, the firm, or both must realize when the service is delivered (this is sometimes referred to as the *value proposition*).

The service concept design should be subject to the scrutiny of an approving body within the firm. The value proposition may be necessary in order to convince the approving body that the idea has merit, such that they are willing to commit the firm's resources to create and test a prototype.

CREATION AND TESTING OF A PROTOTYPE VERSION

Creation of a prototype is the stage where a service idea begins to take shape. The prototype may consist of an explanation of a detailed process that the firm will use to deliver the intended results; or it may include a set of tools and/or techniques that will be required as part of the service delivery process. Once the package of processes, tools, and techniques is assembled, it needs to be tested. A test plan should be prepared that follows a standard testing framework developed by the firm.

The presence of a testing framework is part of the internal discipline and control that is deployed by the firm to ensure that its professionals don't short-circuit the SDP. The prototype should be tested following the guidelines set forth in the firm's testing methodology and satisfy at least the minimum standards to determine the viability of the idea as contained in the prototype. The testing guideline should spell out who might be included in the test audience.

TESTING OF A PILOT VERSION WITH A LIVE CLIENT

When the prototype test results satisfy the established criteria, the idea is ready to be developed into a pilot (production) version. In essence, the firm has become committed to investing significant resources into a delivery system. At the point where the full pilot delivery system is complete, the firm may test the delivery on an actual client. But before this is done, the firm must define specific metrics to measure the effectiveness of the service. These metrics should determine if the service is able to deliver the intended results and benefits to the client and the firm. If the service fails to meet any of the target metrics, the shortcomings should be investigated to determine the cause and whether there is a flaw in the service, the delivery process, or the readiness of the firm's professionals to engage the service. This is necessary to enable the firm to make adjustments to any of the identified areas of weakness and decide whether or not the service is ready for rollout to a target client base.

PACKAGING THE SERVICE

Once the firm has determined that the service is ready to be rolled out to the marketplace, the marketing professionals have to determine how the service should be packaged so that it will command attention and generate interest. Marketing professionals are keen in the practice of differentiating between actual value and perceived value.

An example of this was demonstrated years ago, on a television show called *Art Linkletter's House Party*. Some shows included the segment "Kids Say the Darndest Things," where very young children were featured with Mr. Linkletter, who asked each of them a series of questions. In one episode, three small tree branches were anchored in pots. The first branch had a single $100 bill attached to a branch. A second branch had five $1 bills attached to the branches. The third had twenty very shiny, newly minted copper pennies attached. Each child was asked to pick the tree they wanted, with the understanding that they could keep what was on it. Each child's mother stood by watching nervously, hoping her child would pick the tree with the $100 bill. Alas, child after child picked the tree with the shiny pennies.

The service not only has to be better, but it has to be perceived as *being* better.

PLANNING AND IMPLEMENTING THE MARKET LAUNCH OR ROLLOUT

When the service is ready to be rolled out, firms must take care in planning and then implementing the launch of the service to the marketplace to ensure that the target clients are made aware of the service and its benefits compared with similar services offered by other firms. There is much competition among competing firms for "space," whether it is on a television, computer, tablet, phone, radio, advertising circular, the Internet, or any other communication medium.

A service rollout may occur in several ways. Each must be measured in terms of cost and benefit to your firm:

- Link the rollout to a major trade show or several shows, where it can be featured in your firm's booth and perhaps in presentations to participants. This is a relatively low-cost approach if you already participate in trade shows.
- Use your firm's website, e-letters, and newsletters to announce the service to the marketplace. Except for the cost of mailing newsletters, this also is a low-cost approach to announcing the service.
- Direct mail notifications to current, former, and prospective clients, with follow-up visits to potential clients expressing interest. This approach is a bit more expensive, but inclusion of former and current clients may result in a higher response and engagement "hit" rate.

These are just examples. We recommend that other potential marketing and rollout opportunities should be investigated and tested.

MEASURING SERVICE PERFORMANCE

The firm should also develop specific and measureable milestones to determine if the service has met the objectives for the launch or rollout. The firm needs to know if interest was generated. The volume of positive commentary in the business press or blogs that appear on the Internet by the intended audience may be a valid measure of interest. Sometimes firms are pleasantly surprised to learn that their service far exceeds the performance metrics that were envisioned.

Performance characteristics that exceed expectations may well dictate that a firm should reevaluate its pricing strategy. One of the authors served on a process improvement engagement that had a hard savings target of $20 million. The team achieved $62 million in savings during the first phase of the project. This resulted in the firm renegotiating the pricing for the next phase of the engagement to include being paid a percentage of the savings realized, in addition to the fixed fee. It also resulted in a very satisfied client and an excellent reference.

PLANNING SERVICE GROWTH AND CONTINUOUS IMPROVEMENT

Firms should never rest on their laurels when they are experiencing success. Authors Gerald Nadler and Shozo Hibino refer to the notion of "the solution after next."[1] Nadler and Hibino observe that many good firms are always busy working on their "new and improved version." To be truly competitive, however, a firm needs to anticipate what the competition's next new thing is going to be, and focus on the solution that will come after that—hence "the solution after next."

IDENTIFYING AND CONTACTING POTENTIAL CLIENTS

If you haven't already done so, the first step is to assemble marketing material that identifies the services your firm delivers and the qualifications for doing so. This is an opportunity for the firm to highlight what it does well. The marketing material may begin with a statement that discusses the firm's familiarity with the industry to which it is trying to sell. It may be enhanced by offering a description of the experience the firm has within the industry, as well as prior management experience its service corps may have in the industry. This can be followed by a list of clients it has served, together with the education, background, and experience of the professional service staff. Finally, the materials should highlight those things that differentiate your firm from the competition—those elements that would make your firm the better choice in the eyes of the decision makers.

IDENTIFICATION OF THE TARGET BUYERS OF THE SERVICES

When a description of the services to be performed is refined and the markets and targets are defined, you are now ready to identify candidates that may want to purchase your services. Selling initiatives ranges from making a cold call, at one end of the spectrum, to approaching existing clients, at the other end. It has been said by those with many years of experience providing professional services that it takes ten times as much time, money, and effort to sell to a new client as it does to sell to an existing client. The implicit message is that buyers of a professional service like to buy from someone they know versus someone they don't know. The challenge then is to figure out whom you may know or how to get to know people before trying to sell to them. More specifically, you want to spend your time selling to a decision maker. So you have to figure out who the decision makers are, get to know them, and then try to sell your service to them. The buyer could be a senior executive (e.g., company president or city manager) or a technical/functional director (e.g., vice president of operations or public works director).

Determining Who the Decision Maker Is

Identifying the decision maker is relatively easy. We recommend beginning your research by visiting the client's website, then making a number of exploratory calls to the target client and gaining intelligence on how their procurement process works. We have found that you may encounter a couple of obstacles in this process:

1. In some cases, all proposals must go through the procurement organization. Many organizations have set up governance structures with the announced purpose of establishing a more equitable process of vendor selection. We have found this to be especially the case in organizations that involve the use of taxpayer money. In some organizations, this practice also serves as a way for companies or public agencies to take on the appearance of being fair and equitable. The decision maker and the procurement process administrator are almost always different in governments. Also, service providers may be allowed to prequalify (through a Request for Qualification process) and then be placed on a short list waiting for the need for a related project to arise.
2. Other organizations have engaged in a process that originated in the manufacturing sector as part of the Just-in-Time movement, which seeks to reduce the total number of vendors that the organization does business with. The rationale is that buying organizations should develop partnerships with mutually beneficial goals that will lead to improved quality and productivity.

However, our experience is that executives high enough in an organization usually can heavily *influence* the buying decision, if it has to do with their organization. This experience is reflected in the old saying that "it's not what you know, it's whom you know." For example, it's usually the mayor or their staff (in a mayor/council form of local government), the city manager or county administrator (in a council/manager form of government), a vice president of an operating unit, the chief information officer, or a department head (in a large government entity) that may influence decisions.

How to Gain an Audience with the Target Buyer

There are a myriad of ways to gain an audience with the target buyer. Brad Stribling offers numerous insights to achieve this.[2] One approach is to find a way to get to know the target person and allow them to get to know you. One way is to be introduced by someone whom the buyer knows and trusts. This will require you to do some research. The effort is nothing short of assembling the equivalent of a dossier on the target individual, much like we have all seen done in the movies by secret agents or an elite unit of the FBI pursuing a case. This may include finding out what clubs the buyer belongs to, what interests they have outside of work, determining how they spend their time, and most importantly, who their friends are. Assembling this collection of information has the goal of getting you introduced and known to the target audience.

Other less-direct ways of meeting potential clients are trade shows, industry conventions, professional associations, chambers of commerce, or service clubs like Kiwanis and Rotary. Your strategy will be influenced by your geographic focus (local, regional, national, or international). You will also need to decide if you will be a board member, speaker, exhibitor, and the like, as well as the marketing approaches that you will employ (such as advertising in professional association publications or use of social media). Lastly, you will have to establish a budget and know how much money you may have to spend—national ads and mass mailings can be very expensive.

DEFINING THE PROBLEM THE TARGET WANTS TO SOLVE

You need to do your homework so you are ready when you have achieved the goal of gaining an audience with the target individual. This means that you have to be prepared on a number of levels. First, you will need to have done research on both the industry and the target client. Second, you should gain as much intimate knowledge as possible of the target client's strategic plan. For example, you should know what major issues the industry is facing and, more specifically, how your potential client is positioned to deal with those issues. Finally, you should understand how your service might benefit the client. This requires asking questions and doing a lot of listening. Be aware that an executive's time is considered to be valuable. They will expect when they grant you some of their time that you will have already done your homework, and they won't have to educate you about the issues they are facing.

IDENTIFYING THE COMPETITION

As important as it is to know your service and prepare for your meeting, it is equally important to know who the competition is and everything about them. Moreover, you also need to know what kind of relationship your competition has with the decision maker or a key influencer of the decision maker.

We have come up against competitors in our careers that are well known in the world of decision makers and buyers of consulting services. We have been aware of specific circumstances where we knew we had a better proposal at a cheaper price, which should have made selecting our firm an easy decision. But it didn't happen. We didn't win the bid because we lost to what we characterize as the herd mentality. There was an old saying in our early years of consulting that went something like, "No one ever got fired for selecting IBM." In other words, they selected IBM because everyone else did. The point is that all competitors have weaknesses. Your challenge is to uncover them and exploit them to the max.

There is one other factor that relates to why an inferior competitor was selected instead of us. We call it the Pluto Effect, and it goes like this. After the discovery of the planet Neptune in 1846 and then Uranus, astronomers observed the irregular movement of these planets around the sun. The irregularity could not be explained. Percival Lowell, an astronomer, hypothesized that only another heavenly body of sufficient mass and location could exert a gravitational pull strong enough to cause

Neptune to deviate from its path. In 1930, Clyde Tombaugh calculated the location of this body, pointed his telescope at the coordinates, and discovered Pluto. In business, sometimes bodies move in directions that are contrary to what would seem logical or expected. We've seen unexplained behavior time and again in business. "Pluto" presents himself in many disguises. When we pointed our attention at different coordinates, we always discovered "Pluto." One of us observed a worker on the plant floor erupting in fits of rage when he didn't get his way. We wondered why the behavior was allowed to continue unchallenged. We learned later that he was a key shareholder in the company.

DIFFERENTIATING AND HIGHLIGHTING THE STRENGTHS OF YOUR FIRM

When deciding on the strengths that you should highlight to differentiate your firm from the competition, it is important to first understand the attributes that are valued by executives. According to Barry J. Farber, executives value the following 10 attributes:

1. *Knowledge*—Executives will demand that you have a broad range of knowledge about their industry; significant trends in their marketplace; a thorough understanding of their products, services, and business; and that you know both their customers and their competitors.
2. *Empathy*—You need to be sincerely interested in the client and their business. They want you to demonstrate that their issues are important to you, and they need to feel you understand their unique goals and challenges.
3. *Good organization*—Executives will not waste time with you if you come across as unprepared or disorganized. You need to have done your homework and be prepared for a business-focused conversation.
4. *Responsiveness*—Be accessible at all times and respond to all interactions and requests as quickly as possible.
5. *Appearance*—Executives expect sellers to dress appropriately. They see appearance as a mark of respect for their role and status. See the last section of Chapter 4 for a discussion on dress.
6. *Follow-through*—Executives expect you to make good on your promises, as well as to make them feel that their interactions with you will result in something positive for them and/or the business after the contact.
7. *Punctuality*—Executives expect you to respect their valuable time by behaving accordingly. Being on time for appointments and returning calls or communications shows respect and courtesy.
8. *Hard work*—Executives give 110% and expect you to do the same. They will notice.
9. *Energy*—Everything you do and say should demonstrate high energy and enthusiasm. Executives are always impressed by a positive attitude, enthusiasm, affability, consistency, and flexibility.
10. *Honesty*—The cornerstone in building credibility, honesty demonstrates personal integrity and that you are someone they can depend on for generating results.[3]

An example may highlight how such differentiation might work. One of the authors presented a proposal to a Blue Ribbon Commission working with state officials in a Midwestern state, when the state auditor identified a significant current-year deficit during the first quarter of the fiscal year. The keys to winning the engagement were as follows:

1. Respect for state government (not having a condescending attitude as private-sector specialists)
2. Having an understanding of state government (with examples of similar projects for other states)
3. The ability to respond quickly

This example points out the importance of going the extra mile to differentiate, identify, and highlight the strengths of your firm.

MAKING A PRESENTATION ABOUT YOUR FIRM TO THE CLIENT

Once you have gained an audience with the target client, it is advantageous to rehearse and role-play the actual presentation. Role-playing builds confidence and puts you at ease with the words that are going to be said, such that they roll off the tongue comfortably.

We encourage practitioners to assemble a list of objections that the client is likely to raise, as well as a list of items that you think are weaknesses of your firm. Once the list for each is prepared, the next task is to craft carefully worded language that directly addresses the objections and the perceived weaknesses. Preparation and rehearsal are key. When the potential client comes out of left field with an objection or asks about something that you perceive is a weakness, you will be prepared to answer the questions and able to do so with ease, because you have rehearsed. We believe that these attributes are important, because they allow the professional service firm to lay the groundwork for preparing the proposal when the invitation to bid follows. Our most successful selling experiences have been those where we were so well prepared, that the invitation to submit a proposal amounted to a mere formality.

RESPONDING TO A PROPOSAL OPPORTUNITY

The process that leads to a signed agreement for your firm to deliver services to a client has many twists and turns, often by factors outside your influence. Your goal is to identify as many of these factors in advance, and then manage them to the fullest extent possible.

Once you have convinced the buyer or decision maker to allow you to present a proposal, this is the opportunity for you to shine, to put your best foot forward. The proposal letter is another opportunity to present an image of professionalism, as well as a chance to demonstrate to the client that you were listening and heard what was said. Convincing the decision maker to allow you to present a proposal may not apply in the public sector. Certain requests for proposal in the public sector may highlight certain prequalification hurdles, which you must meet in order to be considered.

Your challenge now is to determine the approach that you will use to conduct the engagement. Previous engagements may be the best source for identifying the phases, activities, and tasks that need to be done and the order in which they should be accomplished. If the option presents itself, you should take the opportunity to assemble an approach description of how you plan to conduct the engagement. Then you can present it to the client as a "road test" to confirm that you are on the right track. Again, this may not be an option in the public sector.

We need to note here that the steps involved in selling to an *existing* client may differ substantially, in that an engagement often comes in the form of add-on work. It has been our experience that the highest vote of confidence one can receive from an existing client is when you are asked to prepare a one-page engagement letter-of-understanding for a significant piece of add-on work. The implicit message is, "We trust you and we have confidence that, while the path forward may not be clear, you will work with us to define the next steps and then act to complete them successfully." Large organizations and the public sector may require a more formal process, including the generation of additional statements of work.

PREPARING AND PRESENTING THE SERVICE PROPOSAL

The proposal should delineate what you are going to do and how you are going to do it, in clear and unambiguous terms (free of buzzwords and euphemisms that leave the readers guessing or inferring what you are going to do for them). Avoid creating a perception that what you are going to do has no value, doesn't differentiate you from the competition, describes services they can do themselves, and/or isn't comprehensive.

The following sections may be included in your proposal document:

- Title page
- Cover letter
- Introduction
- Background to the engagement
- Objectives of the engagement
- Scope of the project
- Deliverables
- Approach
- Work plan

- Schedule
- Progress meetings and reports
- Firm qualifications
- Project team qualifications
- Client participation
- Benefits
- Costs
- Conclusion

Appendix A, Proposal Preparation, discusses each of these sections in detail and includes examples and sample wording. The appendix also includes a discussion of the proposal preparation process.

CLOSING THE SALE

The singular shortcoming that many salespeople cite as the primary contributor to the inability to close a sale is the failure to ask for the order. Their reasoning is that many buyers of services don't know what to do next (of course, if they did know, they would have done it already). Asking for the order drives the conversation in one of two directions. The buyer asks either where they should sign to authorize the proposal or what they have to do to move the process forward. Maintaining momentum is key. You want to control this process as much as you are allowed. For example, if the buyer says that they have to meet with a decision committee, ask if you can attend and make a presentation to that group. Or maintain the momentum by nudging them to establish a follow-up date. Organizational inertia will move the client in the direction of doing nothing. You need to get them to do something.

Another dimension often overlooked during the sales cycle is the opportunity to educate the audience as to the value of your firm's service or the cost of *not* engaging you in terms of missed opportunity. It is often assumed that executives know everything. After all, "He has a master's degree and is a long-time senior executive." You will miss an opportunity if you make the best presentation and then assume that the buyer will make the "logical" decision. You must work to overcome two forces:

1. Organizational inertia will work against making a decision at all.
2. If they are predisposed to making a decision, you don't want to leave your selection to chance.

Many organizations may have an internal champion who will push the process along. Our experience is that this is the exception and not the rule. If you are lucky enough to have a champion, they will appreciate the extra help in working against the propensity for the organization to do nothing. By staying engaged, you can guide the decision making in your direction and manage any misperceptions that may arise. All of these actions will also help to reinforce the client's perception that your firm is one that will keep things moving and will get things done.

ENTERING INTO A CONTRACT

We have seen contracts that range from a single-page document stating that the parties agree to work together and are signed and countersigned, to multipage documents with numerous exceptions and disclaimers. You want to be sure that the contract reflects your proposal (if they are separate documents), regardless of the form your contract takes.

To be sure, contracts require a certain amount of legalese to protect the firm and the client. Firms need to ensure that the contract reflects accurately what was in the proposal. However, wording protection clauses that are totally one-sided does not help when you are trying to establish a trusting relationship. Don't make it too onerous. Too many terms and conditions may infringe on the trust and goodwill that you are trying to build. We recommend that you try to strike a balance. The Mercer Group, Inc., for example, has a standard, lawyer-prepared two-page contract within which the proposal is included by reference. Many times that contract document works fine for the client; other times the client wants to draft the contract.

Author Chester Karrass states that the person who comes to the negotiating table with very clear goals as to where they want to end up usually wins.[4] The contract should specify what is going to be done, who is going to do it, when it will be completed, what constitutes completion, price, and any assumptions on which the outcome is predicated.

WHAT TO DO IF YOUR PROPOSAL ISN'T ACCEPTED

It is a simple fact of life that not every proposal is accepted. We have had our share. Sometimes we knew going in that our proposal was a long shot. We also knew that the more shots you take, the greater the likelihood you will score. But a proposal that has not been accepted may be an opportunity for learning. We have made it a regular practice to solicit feedback from decision makers to learn where there was a perceived weakness in our proposal or what attributes allowed the competition to beat us. Sometimes the feedback is justified, and we may choose to incorporate changes into our presentation or service offering as a consequence.

Then there are those rare exceptions where a follow-up inquiry can change the course of history. One of the authors recounted a case where his firm did not get selected for a large, multiyear engagement—at first. The author was so convinced that his firm was the better selection that he called the key decision maker and asked to schedule a one-on-one meeting so he could plead his case—even though the key decision maker had attended the proposal presentation session, along with his selection committee. As a result of the follow-up meeting, the key decision maker became convinced that his committee had made the wrong decision. After the meeting, he reversed the decision of the committee and selected the author's proposal. We stress that this outcome is probably the exception and not the rule. But it does illustrate that it can happen in certain circumstances.

Our experience, after having worked on the "inside" with clients making selection decisions, is that the process is often flawed:

- Members of a decision or selection committee sometimes miss key meetings because of schedule conflicts.
- Attempts to fill in gaps due to absences may miss subtle nuances or misrepresent the importance of elements of the presentation.
- Decisions can be motivated by hidden agendas and internal politics.

Regardless of the outcome, a professional who walks away from a potential client that did not select their proposal, and then does not ask how he might have done things differently to achieve a winning result, will miss an important learning opportunity.

TIPS FOR SOLE PROPRIETORS AND SMALL FIRMS

Sole proprietors and small firms have special challenges in selling engagements, including time limitations, lack of money for marketing campaigns, and competition from larger firms, perhaps with spiffy looking proposals (e.g., three-ring binders, pictures, tabs) and hordes of people to attend a proposal presentation. Nevertheless, smaller firms have some advantages, including local presence, flexibility, very high commitments to client service, and hunger. Some marketing ideas follow for those who are a small firm or work for one:

1. Develop an "elevator speech." If you can't explain your services in a sixty-second elevator speech, you may need to work on defining what you are selling.
2. Start selling services that are "right up your alley," then branch out as your knowledge and experience grow and determine if growth suits your overall plans for the business. Start with one or two services. Grow to three or four, and then stop, unless your goal is to morph into a much larger firm. More than four services may challenge your ability to manage.
3. Unless you have a partner skilled in other service areas, stick to your knitting. A CPA will conduct financial audits and do taxes as core services. Expanding into other service areas, like information technology and consulting, may undo your core services, unless you are fully prepared to branch out.
4. Grow your practice methodically, pausing now and then for a "How am I doing?" discussion, even if only with yourself.
5. Learn enough about your competition so that you are not surprised in front of a potential client. Proactively cover topics that your competition may point to as weaknesses. For example, a multifirm engagement team might be criticized because they are not all employees of one firm. Counter that you picked this specific team (who are senior and experienced people) for this engagement and you don't need to put low-level staff on the project just to give them some work.
6. Regularly revisit the Ten Client Expectations (knowledge, empathy, good organization, responsiveness, appearance, follow-through, punctuality, hard work, energy, and honesty). Periodically ask yourself two questions:
   - How did we do on the client engagement we just completed?
   - Where do we need to improve for the next client?
7. If you aren't a lawyer, hire one to review *every* proposed contract. Make sure you comply with the details, like insurance, licensing, progress reporting, and billing protocols (e.g., the need to submit copies of expense receipts).
8. Never disrespect a competitor, even if they are disrespectful to you. It's your job to sell yourself and the benefits of your services, not tear down someone else. Learn from each competitor, and prepare counterarguments for competitor objections about you.
9. Always thank the potential client for the opportunity to propose, even if you aren't selected. Being professional in defeat opens up the possibility of future work. Remember, you impressed them with your proposal and presentation on this project, which builds momentum for future work. No sore losers are allowed in successful professional service firms.

ENDNOTES

1. Nadler, Gerald PhD, and Hibino, Shozo PhD. 1998. *Breakthrough Thinking: The Seven Principles of Creative Problem Solving,* Revised Edition. New York: Crown Publishing Group.
2. Stribling, Brad. 2013. *Selling to the Executive Suite.* Carrolton, TX: Selling Executives two-day workshop.
3. Farber, Barry. 1995. *Sales Secrets from Your Customers.* Pages 19–55. New York: The Career Press.
4. Karrass, Chester L. 1992. *The Negotiating Game: How to Get What You Want,* Revised Edition. New York: HarperCollins Publishers, Inc.

4 Setting Up the Engagement

Rules for Setting Up the Engagement

Rule 3: Establish and enforce engagement standards, including those for proposals, progress reports, and deliverables.

Rule 5: Prepare complete and definitive service proposals, contracts, and engagement work plans that evaluate and accommodate risks for both the provider and the client, so the client knows what can be expected in terms of scope, work plan, schedule, deliverables, and cost.

THE ENGAGEMENT TEAM

A number of things need to be done before the firm's engagement team begins working directly with the client. Perhaps the most important part of the engagement occurs during this phase: confirmation of the engagement team leadership and key team members—people who have the right technical skills and personalities to work well with those of key client contacts. The leadership team then develops a detailed work plan, schedule, and budget—as well as a risk mitigation strategy and plan. The communication strategy and plan, both internally and with the client, also are developed and become a part of the work plan and schedule. Many of the engagement set-up activities are a part of the overall engagement control process.

First, it is important that we agree on the definition of a typical organization structure for an engagement, as well as the key participants. Consider the structure shown in Figure 4.1, which reflects a typical professional service engagement. Now let's define some terms:

- *Relationship Leader*—This is the person who has the closest relationship with the key client contacts for the engagement and/or the firm's responsibility for maintaining a relationship with the client. Their role is to monitor and ensure client satisfaction through interactions with client executives and decision makers. The relationship leader is responsible for both the development of new service strategies that meet client needs and the monitoring of the client's perception of the firm's service quality.

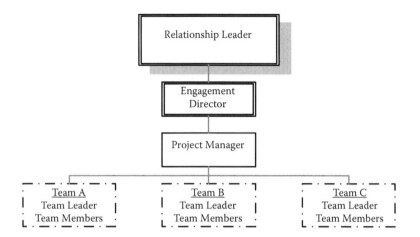

FIGURE 4.1 Typical professional service engagement organization structure.

- *Engagement Director*—This is the person who provides oversight and mentoring of the project manager and ensures the quality of management and engagement deliverables. The engagement director is responsible for ensuring high-quality client service, delivery of appropriate services and products, profitable engagement performance, and properly managed risk on the engagement.
- *Project Manager*—This is the person who supervises team members and manages all aspects of the engagement and deliverables. Some of the project manager's key responsibilities are as follows:
  - Develop the engagement plan
  - Assign tasks to the team leaders and members
  - Act as the key liaison with client personnel assigned and/or related to the engagement
  - Manage and control the engagement on a regular basis
  - Schedule and orchestrate presentations to the client
  - Develop updates to the engagement plan, as the engagement evolves
  - Review all deliverables prior to the engagement director's final approval
  - Prepare client billings for review and approval by the engagement director
  - Ensure that all engagement documentation receives proper handling and security
- *Team Leaders and Members*—These are the people who conduct the technical and/or administrative aspects of the engagement. They perform engagement activities, as assigned and directed by the project manager.
- *Engagement Controller*—An important but less formal role on many engagements is that of engagement controller. The engagement controller is responsible for maintaining up-to-date engagement documentation and for assisting the project manager in executing engagement administration procedures,

such as status control, time management, and client billing. Assignment as the engagement controller may be a first and important step toward assuming broader responsibility for engagement management and leadership.

In a small firm, one person could fill two or more roles; a sole proprietor would fulfill all of the roles, from relationship leader to team member. A very large professional service firm would likely see different people at each level, particularly on large, long-term engagements.

INITIAL MEETING WITH THE CLIENT

Once the client has accepted the firm's proposal, the first step is to meet with the key client contacts to introduce the members of the engagement team—both firm and client. Of course, this introduction may have occurred during the proposal process, but this initial engagement meeting has a number of other purposes. The first is to review the client's overall goals and expectations for the engagement, to ensure that they still conform to the contents of the proposal—or to determine what needs to be modified if they do not. Next it is important to agree upon the general work plan and set key dates for progress meetings with the client, completion of phases or major activities, and presentation and delivery of products. Other initial meeting purposes might include the following:

- Discuss the timing of the engagement and any resource availability constraints, together with confirmation of client staff involvement
- Arrange office accommodations, support services, security, travel, and general administration details with the client, if appropriate
- Schedule an initial presentation to the client's senior management to introduce the project and its objectives, if appropriate
- Determine the method to be used to announce the project within the client's organization, as well as to whom the announcement should be directed
- Collect further information that may not have been available to the engagement team during the pre-proposal and proposal preparation stages

The main objective of the initial client meeting should be to identify and clear up any misunderstandings that may exist regarding the proposed scope of work, approach, timing, cost, and deliverables—*prior* to starting work.

THE ENGAGEMENT CONTROL PROCESS

The next step is to set up the control process for the engagement. Control of the engagement is characterized by seven key elements:

1. Engagement control system
2. Engagement documentation (work papers)
3. Engagement guide (or project definition)

4. Detailed work plan, schedule, and budget
5. Resource loading the engagement
6. Risk mitigation strategy and plan
7. Communications strategy and plan

ENGAGEMENT CONTROL SYSTEM

A firm's engagement control system (ECS) can range from simply a set of hard copy file folders, supplemented by personal computer backup files, to a very sophisticated computerized system with hard copy backup of certain legal documents (such as the signed contract document). In the first case, the firm might be a sole practitioner or very small firm; the second situation would likely be a much larger professional service firm, where a number of people may need to access information in the ECS. Regardless, the types of information that would be contained in the ECS include the following:

- Client name and identifying code
- Engagement name and identifying code
- Key client contacts information
- Firm's engagement leaders and assigned staff members
- Engagement fee and expense criteria
- Time and expense records of professional service and support personnel assigned to the engagement (after charges are authorized)

When the engagement work plan and schedule have been prepared, the related engagement budget can be computed and added to the ECS. The engagement budget may or may not coincide with the negotiated fee and expense total. For example, the firm might wish to "invest" in the client, resulting in a budget that exceeds the contractual fee and expense amount.

ENGAGEMENT DOCUMENTATION (WORK PAPERS)

The work papers are an organized collection of all pertinent engagement information and documentation, both physical and electronic. The work papers typically are created during the proposal development stage and maintained throughout the course of the engagement. The work papers have three very important purposes:

- To combine all important engagement documents and computer files in one place
- To provide an organized method for developing and reviewing conclusions and recommendations
- To enable others to review and monitor engagement performance

Properly arranged work papers are an aid in conducting an engagement, as well as an invaluable source of information when questions arise about fees, conclusions, recommendations, or the conduct of the engagement; when future work for the client is being considered; or when a similar engagement is undertaken for any client.

Substantive support for the deliverables resulting from the engagement is provided in the work papers and made available for review by the relationship leader, the engagement director, and any other reviewing parties within the firm. Such documented support may include the results of fact-finding activities; the analysis of information collected; and alternative solutions, recommendations, and implementation plans developed for the client. The work paper set is made up of two parts: the *engagement file* and the *working file*.

The Engagement File

Typically it is the project manager's responsibility to maintain the engagement file. Initially the file should contain

- All proposal data
- The client-signed engagement proposal and/or contract, including appropriately authorized transmittal sheets
- The engagement work plan and budget

As the engagement proceeds past the set-up phase, the engagement file also might include

- Any amended proposals
- Work-in-process and billing records
- Internal and external correspondence, including notes of telephone conversations and emails
- File notes covering staffing details and general engagement management matters
- An engagement summary
- Relevant analytical data/information used in connection with the work
- Meeting documentation, including attendees, presentation materials, and formal minutes
- Progress reports
- Copies of interim and final deliverables or work products, including appropriately authorized transmittal sheets

Generally the material in the engagement file may be paper based; however, certain analytical material, work products, and deliverables may be maintained in electronic files, as long as they are properly protected and referenced in the work papers.

Sometimes an engagement may be conducted on-site at the client's facility, requiring that the work papers be maintained there. In such cases, key engagement documents should be adequately protected and be replicated off-site and cross-referenced to the records on-site, to ensure accessibility in the event of a dispute with the client. The work papers belong to the professional service firm and should be secured and protected appropriately when they are located outside the firm's office.

The Working File

Each team member working on the engagement should maintain their individual working file, which typically includes, as appropriate

- An engagement diary
- A copy of the proposal or appropriate segment thereof
- A copy of the engagement work plan and any relevant subsidiary plans
- Internal and external correspondence and file notes, including those of telephone conversations and e-mails
- The team member's notes from individual and group meetings
- Meeting documentation, including agendas, presentations, and formal minutes
- Copies of relevant work products
- Other documentation relevant to activities performed by the team member

Again, such documentation may be maintained on electronic files, as long as they are properly protected.

ENGAGEMENT GUIDE

The engagement guide (which often is called a *project definition* document) provides the engagement team with an understanding of the engagement's purpose, intended objectives, expected accomplishments, and key risks. It is a document that bridges the gap between the proposal/contract and the work plan and schedule. The legally governing document for the engagement remains the signed proposal or contract. The engagement guide should not replace or negate any provision of the proposal/contract and should be consistent with those documents.

The development of an engagement guide often begins at the proposal preparation stage and evolves over the life of the engagement as new developments occur. Often it is shared with the client to develop a coordinated understanding of the engagement, particularly the engagement's purpose, objectives, and benefits. A sound engagement guide sets the tone for the engagement's direction, opens channels of communication, sets the basis for developing a trusting relationship with the client and other stakeholders, and tends to bring to the surface any client concerns or challenges. Preparation of the engagement guide typically is the responsibility of the project manager, with review and approval by both the engagement director and the relationship leader. This allows the engagement director to evaluate the project manager's readiness to adequately plan and manage the engagement; a project manager who is unable to develop a clear and concise engagement guide is probably ill-prepared to effectively manage a more detailed work plan, schedule, and budget.

A typical engagement guide contains at least the following components:

- Engagement objectives
- Scope
- Deliverables
- Approach

- Resource and infrastructure requirements
- High-level time and cost estimates
- Key engagement risks
- Risk avoidance and issue resolution strategies

Each of these components is described in the following paragraphs.

Engagement Objectives

Statements of engagement objectives should be specific, measurable, and provide a time frame for accomplishment. Good objectives state what is to be achieved and/or the results sought, not how the engagement team will get there. They help clarify the client's and other stakeholders' expectations. They also help to keep the team focused as the engagement progresses and establish a means for assessing success at the completion of the engagement.

Scope

The statement of scope sets the boundaries for the engagement. It clarifies what will be included in the engagement and, if necessary, specifically what will not be included. The scope statement must be clear, concise, and complete, as it will serve as the basis for determining if and when out-of-scope work is being accomplished during engagement execution—which likely will result in unsatisfactory engagement financial results, due to budget overruns. Potential out-of-scope work should be identified *before* it is performed, so it can be negotiated with the client as additional effort, along with its attendant cost and schedule implications.

Deliverables

A deliverable is anything produced on an engagement that supports achievement of the engagement's objectives. It is any measurable, tangible, and verifiable outcome, result, or item that is produced to complete a task, activity, or phase of an engagement. The term *deliverable* is often used to refer to an engagement product that is handed over to the client and subject to their review and approval. Appendix B, Deliverable Preparation Guide, provides tips on preparing and presenting deliverables, whether a short letter with attached documents, plans, and drawings, or a longer technical report in one or more volumes.

Approach

The approach sets out the general course of action that will be taken to accomplish the engagement's objectives. It may be defined in such terms as the methodology to be used, the timing and phases of the engagement, and/or the types of technology and human resources to be employed. The approach explains, in general, how the engagement will be carried out.

Resource and Infrastructure Requirements

Resource and infrastructure requirements for professional service projects typically fall into any of three categories:

- Human resources
- Facilities and equipment
- Information technology

Human resources, including professionals and support positions, often are the major cost factor in a professional service engagement. In the engagement guide, they typically are defined in terms of the roles, responsibilities, and skills that will be needed for the project to be successful.

Facilities and equipment includes physical office space, both at the firm and at the client, as well as the noncomputer office equipment the firm needs to successfully deliver its services, perhaps including vehicles.

Information technology includes office and personal computers, printers, smartphones, and other technology needed to deliver professional services. Key professional service technologies include the following:

- Effective communication systems (internally and with the client)
- File and data transfer capabilities
- Deliverable sharing capabilities
- Presentation software
- Engagement management system

High-Level Time and Cost Estimates

High-level time and cost estimates are used to gauge and validate the size of the engagement. These estimates set out the management and staffing levels and related costs that will be necessary for major engagement activities, as well as the elapsed time for each activity. Such estimates may be based on experience with other, similar engagements conducted by the firm or from any work sessions that may have been conducted during preparation of the proposal and the engagement guide.

Key Engagement Risks

An engagement risk is any factor that challenges the engagement team's ability to manage the budget, resources, time, and/or quality of the engagement deliverables, as well as acceptance of the deliverables by the client. Key risks are those uncertainties or vulnerabilities that could have a major impact on successful completion of the engagement. We discuss later in this chapter how to go about identifying engagement risks and ranking them to arrive at a list of key risks.

DETAILED WORK PLAN, SCHEDULE, AND BUDGET

A proposal that has given rise to an engagement typically contains a work plan and schedule with sufficient detail to obtain the client's agreement to proceed with the work. But the proposal work plan and schedule may not be useful as a project planning and control tool when the time comes to conduct the engagement. Here are some of the reasons that might give rise to this situation:

- The level of detail in the proposal's work plan may be insufficient
- The proposal's work plan might not reflect the client's comments or new facts that may have come to light since the proposal was prepared
- All necessary sources of information may not yet have been identified or available
- The definition of the methodology to be used may need further refinement
- The time schedule shown in the proposal may have been tentative and needs to be updated
- The engagement organization and staffing indicated in the proposal may have been tentative and needs to be confirmed before the project can start, including any role that client staff may have in support of the work of the firm (e.g., general data collection, financial analyses or reports, equipment and facilities inventories)

Some of these "deficiencies" may have been addressed in the initial meeting with the client (discussed earlier in this chapter). If not, they should be addressed during the engagement work planning stage.

Finalization of the work plan and budget, following the initial client meeting, is central to effective management and control of the engagement and to ensuring that quality service is provided to the client. The nature and scope of the work plan may vary, depending on the size and scope of the individual engagement. Regardless, any work plan should

- Be based on clearly defined engagement objectives
- Be separated into tasks that are manageable units with tangible outputs or outcomes (which may include deliverables)
- Reflect the interrelationships of tasks, and of task outputs and outcomes
- Define the technical and quality review approach
- Assign responsibilities
- Communicate to the team members what is to be produced, when it is due, and the budget allowed
- Be clear enough for review and interpretation by others who are not directly involved in the engagement
- Enable the project manager and others concerned to anticipate and react to future events
- Provide the framework for measuring and reporting progress and generating progress billings to the client
- Provide the ability to measure actual progress in the same dimensions in which the engagement was initially planned and upon which time and fee estimates were based
- Serve as a permanent record of what was expected in terms of tasks, responsibilities, deliverables, time, and cost

To serve as a useful management tool, the project plan and budget details the engagement's tasks; assigns tasks and hours to engagement team members, including client personnel and any subcontractors; generates the budgeted cost of each task; compares actual time spent to the budgeted amount and gives an estimate for completion of each task; calculates total engagement cost, including both fees and expenses; and provides an estimate of the cost to complete the engagement. A number of computer packages are available on the market that will satisfy these requirements, in the case that a professional service firm does not already have the capability in-house. Figure 6.1 in Chapter 6 provides an example of a simple budget and billing worksheet used by one of the authors.

Once completed, the refined work plan and budget should be reviewed and approved by the engagement director. This review may require reconciliation with the timing and cost estimates given in the accepted proposal (or engagement contract), if the two work plans differ significantly. If the refined work plan yields a higher cost estimate or a later completion date than stated in the proposal or contract, then it will be necessary to assess the size of the discrepancy, the firm's related risk, and the causes of and justification for the discrepancy. If the issue is cost, the following options might be considered:

1. If the engagement team is convinced that engagement quality would suffer with a lower level of effort and the firm–client relationship permits, advise the client of the situation and seek approval to proceed with an increased fee.
2. Divide the engagement into discrete elements and obtain the client's agreement to proceed only with an initial portion of the project.
3. Redesign the work plan to reduce the overall cost; options include
   a. Eliminate marginal activities
   b. Reduce the depth of any analytical steps in the work plan and inform the client of cost/quality trade-offs
   c. Rely more heavily on properly qualified client personnel to conduct the work, but only if this will actually reduce the firm's resource requirements.
4. Renegotiate the engagement scope with the client, if the firm/client relationship permits.
5. Accept a lower profit level on the engagement.
6. Resign the engagement, which likely would have a significant adverse affect on the firm/client relationship for years.

If the issue is timing, and there is no possibility of advancing the completion date with the staffing planned for the engagement, the options are as follows:

1. Advise the client and obtain approval to proceed (point out when preliminary or intermediate results will become available).
2. Enlarge the engagement team (which might result in increased project costs).
3. Ask engagement team members to expand their working week, as necessary, to meet the timing objectives.

Whether the issue is cost or timing, a combination of the options may be used. The engagement director, in consultation with the relationship leader and the project manager, is ultimately responsible for the decision on how to proceed. This may require the submission of an amended proposal and/or a contract amendment.

RESOURCE LOADING THE ENGAGEMENT

Without question, appropriate and technically proficient persons should staff all engagements. One of the engagement director's responsibilities should be to ensure that all engagement team members are

- Qualified by training and experience to conduct the activities assigned to them
- Provided with appropriate methodologies and/or tools to do their work
- Allowed adequate time to carry out their assignments
- Adequately supervised on a day-to-day basis

An engagement should not be performed until appropriately qualified and experienced staff members are assigned. This highlights the need to ensure that the qualifications of *client* personnel who might be assigned to the engagement are matched to the requirements. Qualifying the engagement staff may require their participation in special training regarding the methodologies to be employed. Such training could be in the classroom, by self-study, and/or by computer-based learning. Client personnel and subcontractors who are expected to participate as part of the engagement team may also need to be trained.

The engagement team members also should be exposed to appropriate background material about the client, its industry, and the particular situation it is facing. Sometimes it's best to prepare a booklet of relevant information (hard copy or digital), so all team members are aware of the client and its needs.

RISK MITIGATION STRATEGY AND PLAN

Risk is an element that is inherent in every service delivery engagement, regardless of industry. Risks are simply the things that could cause an engagement *not* to go as planned. This is sometimes referred to as Murphy's Law: "If anything can go wrong it will, and usually at the worst possible moment." Again, let's establish some definitions. We say that a *risk* is something that could occur, but hasn't yet. On the other hand, an *issue* is a risk that has occurred.

Risk mitigation is made up of two components: strategy and planning. Risk is, in fact, something that can be managed. The secret is to use the process often referred to as 20/20 hindsight, but applying it as foresight. Risk can never be completely eliminated, but it can be managed. The degree to which it can be managed is directly related to the probability of a project being completed on time and within budget. Managing risk means that fewer unexpected events will occur, and for those that do occur, a contingency plan is already in place to deal with the issue.

Developing the Risk Mitigation Strategy

The first step in the risk mitigation strategy process is to identify possible risks. This means that we have to engage in an intentional process of identifying all the possible things that could go wrong. We use the term *intentional* because the risk identification process requires that we not go through the motions so we can say it's done. The process requires the participants to really think outside the box. We have seen significant success with the following approach.

Key stakeholders (both firm and client) are identified and assembled, with the specific objective of conducting a brainstorming session. The purpose of the session is to identify all the possible things that could go wrong. One of the benefits of these types of sessions is that sometimes some pretty outlandish scenarios are offered, and that is exactly what you want to have happen. A key dynamic of brainstorming is that each idea triggers another idea, by individual association and experience.

The person who leads the brainstorming session should be a trained facilitator. A particular brainstorming facilitation skill is the ability to draw out a quiet participant who might not otherwise contribute to the group process.

But how does one know where to start looking for ideas? One approach is to start with design criteria or standard policies and procedures, then combine different pieces into unlikely scenarios. An example of an outside-the-box scenario would be to take different, seemingly unrelated events and combine them to see what impact they may have together. Japanese nuclear power plant planners considered the harm that could be caused by an earthquake, by a tsunami, and by a typhoon—but they did not consider two or three of these events happening at the same time. The result was nothing less than catastrophic, when one considers the impact resulting from the meltdown of several of the reactors at the Fukushima Daiichi Nuclear Power Plant in 2010. Using the brainstorming approach outlined earlier, if stakeholders had looked at the design criteria and seen that planners had included a backup generator in case of a power failure, they might have asked: "What if the generator wasn't working?" or "What if it was underwater?" Like a three-year-old child, keep asking What If?… What If?… What If?… until you run out of What Ifs.

Once the brainstorming team has identified all the possibilities they can imagine, the next step is to rank the probability of each risk's occurrence on a scale of 1 to 3, with 3 being high probability and 1 being low probability. The entire team needs to participate in the process of determining the probability of a risk occurring. They need to deploy the same level of imagination in trying to determine probability. Knowledge of similar scenarios may add credibility to a given probability argument.

Next, each risk is ranked in terms of the impact the risk would have on the engagement or the client if it became an issue (3 would be high impact and 1 would be low impact). The numerical impact ranking assigned to each identified risk may be determined by assigning a dollar range for each value, a numeric value based on the impact the risk may have on the client's strategic plan if it materialized, or a numeric value based on the harm to the client's brand that may result. The following items should be considered at a minimum when assigning weights to the supposed impacts of each risk:

- A statement of the issue—a description of the risk if it were to materialize.
- An impact statement—a detailed statement of the impact the issue would have if it were to materialize.
- The cost of the impact—stated in financial terms, as well as its impact on people and/or property. The cost could also contain an assessment of the impact on nonphysical items, such as the project schedule or the political status quo. What will be the impact if the project is delayed? How will the public at large view the client?
- The cost of mitigation—stated in financial terms and/or the hardship that would be avoided by exercising the mitigation.

When complete, the risk probability ranking is multiplied by the risk impact ranking, to come up with a weighted score for each risk. All of the risks are then sorted in descending value order, with the risks with the highest numerical value at the top of the list. These are the so-called show stoppers.

Thus ends the strategy development portion of risk mitigation. The team has identified all of the risks that it can think of. The higher the numerical risk value, the greater the amount of attention that needs to be paid and effort expended in formulating a plan to prevent the risk from occurring—and having an iron-clad backup plan to execute in case the risk materializes despite the best efforts to prevent it. The next step is to decide what to do about the high-value risks.

Developing the Risk Mitigation Plan

Risk mitigation planning is the process of preparing and executing a plan or plans designed specifically to prevent the risk from occurring and becoming an issue. This includes having a backup plan in case the risk *does* become an issue—despite your best efforts to prevent it.

The first half of the process is to develop a set of actions that the team will take to prevent a risk from becoming an issue. Not surprisingly, often a risk mitigation plan includes steps within the firm to ensure that internal disciplines and controls are in place and are followed.

The second half of the process is to design a backup plan, which spells out exactly what you are going to do if, despite your best efforts to execute the risk mitigation plan, the risk becomes an issue. Having a backup plan has shown, time and again, the value of preparing for all possibilities. For example, the Deepwater Horizon 2010 oil well explosion in the Gulf of Mexico resulted in an estimated 53,000 barrels of oil per day being released into the ocean for three months, while engineers worked around the clock to design and build an emergency shutoff valve. But the time to design and build a backup shutoff valve for an oil well that is thousands of feet under the sea is not after the oil platform has exploded. A risk mitigation plan should have included steps to ensure that all safety procedures were followed, including steps to be taken if people decide to cut corners because they are behind schedule. The plan also should have included the design and construction of an actual emergency shutoff valve that was ready to deploy. Notwithstanding the cost, it may have alleviated some of the damage done from this incident.

Many projects don't proceed as planned, due to the occurrence of unplanned events. The goal here is to reduce the number of unplanned events to as near zero as possible. The risk mitigation plan should have a detailed plan constructed for each risk that is deemed to require one. The plan should contain two parts at a minimum:

- The detailed steps that will be taken to prevent the risk from materializing
- The detailed steps that will be taken in the event the risk materializes

COMMUNICATIONS STRATEGY AND PLAN

Tom Peters (*In Search of Excellence*)[1] told a wonderful story about a steel company that was growing by leaps and bounds, while the competition was closing plants.

Peters made an appointment to visit and rushed to the company to meet the CEO. He ran into the lobby, where the CEO was waiting for him, shook his hand vigorously, and asked, "Tell me your secret! How did you do it?" The answer came back, "It's simple. Our people *talk* to each other." And there you have it: Communication is one of the keys to success.

A communication strategy simply states *what* you are going to communicate, while a communication plan states *how* you are going to communicate. Actually it's *not* as simple as that. A communications strategy and plan has to be thought out well, executed well, and then checked to be sure that the message you thought you were sending is the message that was received. The communications strategy and planning team needs to focus on the key goals of the engagement. Hopefully these have been crafted and are clear to the team.

Developing the Communication Strategy

A well-constructed communications strategy should answer eight questions, which are shown in Table 4.1. Each of these questions is explored in greater detail in the following paragraphs.

1. *What is the message to be delivered?* The first step in developing a communication strategy is to define each message that you want to communicate. This can be accomplished by focusing on the key goals of the service to be delivered—which may involve a number of components, depending on the nature or scope of the service. It may help to consider not only the message you want to communicate, but also the behavioral outcome that you expect to see as a result of having delivered the service. Defining the behavioral outcome leads you to get very crisp in crafting the message. One also has to focus on the frequency of the message, as well as what other messages may be broadcast adjacent to it, to prevent the message from getting lost in the clutter.

2. *Who is the audience?* The intended audience may be internal to the organization of the professional delivering the service, or it could be an external client. The choice of wording and even the vehicle used to transmit the message will vary, depending on the intended audience. In any case, you should

TABLE 4.1
Communication Strategy Questions

1. What is the message to be delivered?
2. Who is the audience?
3. What roles and responsibilities will be impacted as part of the engagement?
4. What is the purpose of the message?
5. How will the message be delivered (what channels, media, and/or venues will be used to communicate the message), and who will deliver it?
6. Who will prepare the message?
7. What are the timing and frequency of each message?
8. What feedback mechanism will be used to ensure the intended audience received the intended message?

be specific about who the intended audience is. The message that needs to be delivered to senior management members may be different from the message delivered to the rank and file. The difference in messages is not related to the position one holds in an organization; rather, they need to be aligned with the job function being performed. The makeup of the audience may require multiple iterations or multiple levels. The intended audience may include senior management, middle management, and/or entry-level personnel or trainees. Each of these groups may require slightly different to major differences in each of the elements of the plan.

3. *What roles and responsibilities will be impacted as part of the engagement?* Identifying the roles and responsibilities to be impacted is related to who the audience is. This is an added dimension that will help define how an initiative may impact the members of the target audience, as well as the changes in behavior that are expected as an outcome of the delivered service.

4. *What is the purpose of the message?* Defining what the message is and determining the purpose of the message are closely linked. Focusing on the purpose of the message will help you bring a sharper focus on defining the message that you want to communicate.

5. *How will the message be delivered (what channels, media, and/or venues will be used to communicate the message), and who will deliver it?* Invention of social media has revolutionized the avenues available to communicate a message. Further, the dynamic nature of social media, that it can be real time *and* interactive, provides an additional perspective in crafting the message to be communicated. The dynamic aspect of social media is not limited to the interactive dimension. The dynamism relates to the rate of technology evolution as well. While some are just learning to use e-mail, others have moved to cell phones, while others have moved to smartphones, where texting has relegated e-mail passé. Now tablets have upstaged these relatively new technologies. That begs the question of by whom and in what way will the message be delivered (i.e., to what extent does the intended audience have the technology necessary to receive the message from the sender?).

6. *Who will prepare the message?* Who the message will be delivered to and/or who the intended audience is may determine who will prepare the message. The creator of the message needs to be able to "speak the language" of the target audience and be conversant in the medium or technology to be used.

7. *What are the timing and frequency of each message?* A message that will appear on an easel in the lobby of a firm or in the firm's monthly newsletter is very different from an ongoing dialogue on a blog or in a series of "tweets." Use of the latter option needs to be crafted carefully; a good message or a mistake will be promulgated at the speed of light.

8. *What feedback mechanism will be used to ensure that the intended audience received the intended message?* You need to be able to confirm that the message you sent is the message that was received. One way to do this is to develop a *traceability matrix*, which links a particular message and its transmission vehicle to an expected behavioral outcome. Questionnaires, scorecards, and dashboards may also be used to track responses.

Developing the Communication Plan

The answers to the eight questions define the communication strategy (i.e., what we are going to do). The communication plan, on the other hand, can be constructed using a simple spreadsheet template. Figure 4.2 is one format that we have found to be useful in assembling a communication plan. This format not only serves as a vehicle to assemble answers to all of the strategy questions, but also allows one to capture the answers for each of the audiences that have been identified in communication strategy question number two. A spreadsheet template could be designed, using such a format, with tabs at the bottom to separate the distinct audiences.

Table 4.2 provides some examples of the types of communications that you may want to consider during the development of a communication plan for a professional service engagement.

CONDUCTING FURTHER RESEARCH

While the engagement control process is being established, it may be necessary to conduct further research on the client, the nature of the engagement, and/or the methodology to be used. Some research may have been conducted during the engagement selling stage, to provide background for preparation of the proposal. But typically this is not enough after the engagement has been obtained. In particular, preparation of a detailed work plan, schedule, and budget may require research into, for example, the alternative methodologies that might be employed, the number of people who might need to be interviewed, or how long it might take to conduct certain engagement activities. The risk mitigation planning process may require investigation of alternative scenarios and the experience of other related engagements—either within the firm or with competitor firms (particularly through use of the Internet). Regardless, it is important to recognize that research does not end with submission of the proposal.

DEVELOPING A LIST OF ENGAGEMENT ORIENTATION INTERVIEWS

The engagement's communication plan should be the starting point for developing a list of engagement orientation interviews. These interviews provide the basis for gaining a better understanding of the client's desires, issues, and any constraints that may be placed on the engagement team. They may also help isolate any unidentified risks that need to be incorporated in the risk mitigation plan. The orientation interviews also provide the opportunity to develop a positive relationship with client personnel, as well as inform them about the firm and the purpose of the engagement.

The orientation interview list typically includes those client people who will be directly involved in the engagement, as well as those who might be affected by the engagement results. Those firm members who will be conducting the interviews should be skilled interviewers and fully prepared, with checklists of what to ask and space to record the answers. Normally the client orientation interviews are conducted as one of the early steps in the engagement work plan.

ABC Company		REMARKS/BACKGROUND
MESSAGE: ABC Company is a company which promotes: • **Great standardization and accountability in Training** • **Formal certification in the Technical and Leadership aspects of the job** • **Increased interface with upper management during the first few years of employment** • **Greater opportunity for more rapid career growth and progress** • **Improved workforce planning, keeping the pipeline full of well trained and educated managers, capable of taking on increasing levels of responsibility**		
Audience (Who needs to know?)	Senior Management Positions	
Delivered By	Name of person	
The Message (What are we doing and what they need to know)	To be determined	
Purpose for the Message (Why are we doing this?)	A. Position ABC Company as a way to standardize and measure Management Training & Development across the company B. Present ABC Company as a professional service delivery company to help address the challenges posed by the anticipated loss of experienced managers through retirement C. Emphasize management's responsibility to identify, grow and develop new leaders for middle management levels	

FIGURE 4.2 Communications plan template.

Anticipated Roles and Responsibilities	A. GMs will act as sponsors for the trainees during the Trainee's initial 1–2 years of their time at the company—shepherding the trainees and providing periodic counseling. B. Managers will conduct a semi-annual meeting to review current and recent management trainee progress and identify high potential talent and potential developmental assignments. C. Managers will be designated as executive level sponsors for high potentials.	
Prepared By	Senior Management HR and Communications Dept.	
Channels/Media/Venue (How will they get the message?)	A. PowerPoint presentation and FAQ sheet B. Roundtable sessions C. Online university	
Timing/Frequency (When and how often will the message be delivered?)	A. Upon hire	
Feedback Mechanism (How will we know they got the message?)	A. Traceability Matrix B. Trainee and Scorecards/Dashboards	
Date Message Developed	TBD	
Date Message Delivered	mm/dd/yyyy	

FIGURE 4.2 (continued) Communications plan template.

TABLE 4.2
Sample Communication Plan Elements

1. An introductory letter to all affected employees notifying them that a project has been started
2. An introductory letter that briefs these employees on the nature of the project, the process to complete the project, and their involvement
3. Notification of a kickoff meeting
4. Information needed to complete the project
5. List of suggested meetings and/or focus groups
6. Progress/status reports, which may include preliminary deliverables, issues, and ideas
7. Biweekly or weekly conference calls
8. Public announcement to the media (used if public/community forums are scheduled as a means to boost participation)
9. Schedule and general structure of draft deliverables for the service(s) to be provided
10. Schedule of the final deliverable
11. Schedule and structure for final presentations to the governing board, departments/agencies in the scope of the study, and perhaps citizens in a public meeting
12. Interviews with media outlets about the project and/or results, as needed and appropriate

PREPARATION OF AN INTRODUCTORY LETTER

Often it is useful to prepare an introductory letter that provides the firm's name and background, as well as the nature of the engagement. The letter may also include the following:

- The specific objectives to be served by the engagement
- The nature of the deliverables and their timing (schedule)
- The client personnel or groups that are expected to be involved

We do not recommend that the engagement's expected cost be included in the letter, although this information may be well known (e.g., in a government contracting situation).

Although members of the engagement team may draft the introductory letter, it is preferable that a client executive should release the document under his or her name. That will tend to lend more weight to the letter and show that the client is fully behind the engagement. Normally the client should issue the letter prior to the time that the engagement team begins their engagement orientation interviews.

GETTING ORGANIZED TO SHOW UP

If the work is to be conducted at the client's location, a number of logistical aspects must be addressed before the engagement team arrives, such as the following:

- Date and time of the arrival
- Transportation to and from the client's facility, as well as housing for the team members, if necessary
- Specific location of the facilities

Another consideration is the client's dress code. Some years ago, almost all professional service firms required their personnel to wear a suit and tie (or equivalent for the ladies). But now, with the relaxation of dress codes, it's difficult for professionals to determine what's appropriate when visiting a client. As a result, it's best to inquire what the client's dress code is. Barry Mundt does so, and then dresses at least one step above—up to a suit and tie. That way he doesn't appear overdressed or underdressed in the eyes of the client personnel.

TIPS FOR SOLE PROPRIETORS AND SMALL FIRMS

You have put a lot of effort into winning the engagement and might be really tired. Now, in a few days, you've got to rise up and actually do some work, which is a separate challenge altogether. Small firms need to be particularly attentive to the following suggestions:

1. Decide who does what on the engagement when you propose; it's particularly important to define key leadership and technical positions. Then make sure the client knows you are fully committed to the proposed team for the engagement. No last-minute swaps, unless there is a really good reason (like someone quit!).
2. Collect client- and engagement-related information before you go on site to kick off the engagement. Read it through and be fully prepared for that first big meeting.
3. Start with a simple *engagement control system*; then expand and mature it as you gain experience. A simple engagement file using three-tab file folders, for example, might be set up as follows:
   - *Left Tabs*: Project Management Information files, such as
     - Project Management (budget, schedule, work plan, notes)
     - Request for Proposal (if there is one) and Proposal (and amendments)
     - Contract (and amendments)
     - Internal and Client Status Reports
     - Bills
   - *Middle Tabs*: Fact Finding Information files, such as
     - General Information
     - Information about the Client
     - Engagement Research
     - Notes from interviews and focus groups
   - *Right Tabs*: Ideas, Analyses, and Report files, such as
     - Emerging Issues
     - Analyses
     - Draft Deliverables (with client review notes)
     - Final Deliverable
4. Identify risks as early as you can (pre-proposal is best); then create and continuously update the engagement risk mitigation plan. Make your risk mitigation plan more than a mental "note to self." Some common risks to plan for are as follows:

- Changes in key client staff, including the decision maker who selected you.
- Changes in the firm's staff due to illness, life event, moving to another firm, poor performance, etc. It is imperative to have an immediate backup for the project manager (likely the engagement director). This is a huge issue for sole proprietors and very small firms.
- Upheavals in the client organization (e.g., sudden product recall, loss of financing, production breakdowns, failure of IT system, negative Dun & Bradstreet report, new officials after an election for a government, board changes in nonprofits).

5. Develop a standard communications plan for your services and imbed the plans in your standard proposal document, so the plan is planted in the client's mind before you start the project. Clients tend to want assurance that a professional service provider will communicate very effectively before they select one to do a project. Key areas to set up communication plans for are as follows:
   - Initial information request (budgets, financial reports, contracts, etc.)
   - Model employee briefing letter
   - List and sequence of interviews
   - Formal progress/status reports

ENDNOTE

1. Peters, Thomas J. and Waterman, Robert H. Jr. 1982. *In Search of Excellence.* Harper & Row. New York.

5 Adapting the Engagement Plan to the Real World

Rules for Adapting to the Real World

Rule 6: Be flexible and adaptive to the "real world" when the project starts, to manage the dynamic between client expectations and what's really happening within both the firm and the client.

Three aspects that have to do with adapting the engagement plan to the real world are discussed in this chapter:

- Internal versus external client
- Cultural accommodation
- Rapid deployment

The first is the nuance associated with being an internal professional (being an employee of the "client") or an external professional (a contractor to the client); the way one goes about managing the relationships with the organization's other employees may differ. The second has to do with accommodating the culture that exists, relative to the situation at hand. The third issue discussed is a relatively new paradigm for project timelines: rapid deployment. A key question that is explored is how one adheres to internal discipline and control, which arguably takes time, when the client wants to get things done rapidly.

INTERNAL VERSUS EXTERNAL CLIENT

Sometimes service delivery professionals may not have clients who are external to their firm. Some organizations maintain an internal professional service delivery unit in-house. Often, such organizations will entertain proposals from outside firms to foster competition, as a way to drive costs lower, to motivate professionals to improve their skills, or because they have become less than satisfied with the performance or results of the internal professional organization's work. Similarly, sometimes the client may decide to engage your firm to perform duties in which your firm appears to be part of an internal unit in the organization, when in fact you are not. Special care may need to be taken, depending on the role that you have been asked to play. Assumption of a staff augmentation role may require you to blend in. In contrast, an engagement that asks you to perform an objective assessment may require you to stay at arm's length from client personnel and even assume a detached demeanor.

The role of the service delivery professional may require special consideration depending on who the client is. Much has been written and said about whom one thinks the client is when performing an engagement. Some argue that the client is whoever writes the check. Others argue that it is the client's customers. The answer is both of the above. When a professional is hired to deliver a service to a client, they need to be mindful of (1) who the client's customers are and (2) how delivering their service impacts the perception the customer may have of the client, while the service is being delivered.

One must be mindful of the client's internal politics. A professional who is working as an employee may have to contend with forces, real or imagined, that have the potential to influence how they approach a situation. This is particularly the case if they judge that their recommendations or response may impact their longevity within the organization. Such a conflict of interest is ample reason to use an external professional to perform the work.

For example, one of the authors was engaged as a member of a team of internal consultants that was charged with implementing major process improvements. The scope of the project included a review of the efficiency of the client's customer service team and how they interact with customers. One of the recommendations resulted in a change in which customer service agents served the next customer in line. Some customers had their favorite person with whom they preferred to interact when executing transactions. The author had to weigh carefully the impact a potential process change might have, how it might be perceived by the client's customers, and if a change might cause some customers to take their business elsewhere.

Another consideration is the length of time a professional spends at a client's site, which may result in a "tipping point" being reached. Continuing to work beyond the tipping point may cause client employees' to change their perception of the professional. There may be danger associated with staying too long; the line that defines whether the professional is an employee or a contractor may become blurred. The danger is that the professional's objectivity may become compromised or may be perceived as being compromised. This is especially important if the client is paying a premium for the professional's arm's-length and unbiased objectivity. This dynamic may be further complicated if the professional is a former employee. For example, Steve Egan returned to Fulton County, Georgia, to do a series of consulting studies about six years after he had left to become a consultant. People welcomed him back as though he were rejoining the county's staff. At morning break time, he went to the cafeteria and noticed that the same people were sitting in the same seats and eating the same thing as six years earlier. He had to work hard to emphasize he was back as a consultant and to assert his independence and objectivity.

CULTURAL ACCOMMODATION

Cultural accommodation refers to the notion that one must not be wedded to one's own methodology. We have witnessed numerous instances over the years where project managers have been unwilling (or unable) to adapt their project management methodology and process to the culture of an organization or to fast-moving and/or unanticipated events that arise. Following are some examples that have surfaced over the years, as we prepared to deliver professional services.

FEES AND THE PERCEPTION OF VALUE

Early in his life as a consultant, one of the authors was having difficulty articulating the cost of an engagement to a prospective client. His previous experience had never exposed him to the orders of magnitude reflected in the fees being charged, as they were multiples of what he had earned on an annual basis. Further, he did not have an appreciation for the value of the service that he was delivering (i.e., the value proposition). He presented his dilemma to a partner in the firm, who explained the notion of value to him. First, the partner counseled him that he needed to stop thinking like an hourly worker. Knowing that he had experience as an inventory manager, the partner asked him if he could save a company $1 million through better inventory management practices. The consultant said he could. The partner then asked him, if he were a manager at the client company, how much he would be willing to spend to achieve that kind of improvement. Would he be willing to spend $100,000? The consultant answered that $100,000 would be a bargain. He added, "A 10:1 return on a $100,000 investment would be a no-brainer." The focus is changed from charging fees based on the amount of time spent working on the engagement to the outcome achieved—saving the client $100,000, while being mindful that you also want to maximize the fees realized. The partner had introduced the notion of *value* to the consultant.

The consultant went on to recognize that there is more than one dimension to the perception of value; it can be a relative thing. The following illustrates this concept. The consultant was part of a team that was asked to prepare a proposal to help a company, which was an audit client of the firm, to implement a Just-in-Time system at its factory. The team performed its customary assessment by gathering facts and then preparing and submitting its proposal. The proposal was for $120,000 and included an additional 20 percent for contingency. A competing firm submitted a proposal for more than $1 million. But the client selected the proposal from the competing firm. The client's CFO explained that they had done so because it was obvious by the dollar amount of our firm's proposal that there was a lot that we did not understand about a Just-in-Time environment. We knew that the competing firm had a practice of hiring many inexperienced consultants, putting them on projects, and training them on the client's nickel. But none of that mattered.

Thus, the author had a second big lesson on the perception of value. The price one attaches to a proposal is also a statement about the perception of the value of the service to be delivered. Subsequent years of working as a management consultant for different firms confirmed this fact. A number of firms offer to perform analysis work at cost as the first phase of a multiphase effort. The problem is that the up-front price that a firm charges sets the expectation for future phases. The other is that clients are less vigilant about making key resources available for you to perform your services. They reason is that if key stakeholders skip a meeting because they have "more important" things to do, the actual cost to them is insignificant. Conversely, substantial fees will often motivate managers to ensure that stakeholders make themselves available.

NOTION OF INDEPENDENCE

The notion of independence may take on more than one dimension. It may refer to an objective, unencumbered perspective, in which the future of the professional making the observations is not affected by the outcome. It may also refer to a scenario where the professional is able to operate independently, free from the day-to-day internal responsibilities of a position within the firm.

Objective, Unencumbered Perspective

The first kind of independence provides the professional with

- the time to analyze and question the behavior that is observed;
- time to analyze issues that are causing "pain" for management; and
- relative freedom from political repercussions, because the service professional does not have "turf" that they are trying to protect or expand.

An example where the independence of the professional may be compromised is when it is at the behest of a client. We have witnessed situations where a decision maker has chosen to engage an outside professional to create the perception of independence. For example, a client executive may have staked out a position that he would like the professional to recommend. The executive then engages an outside firm to recommend the position independently and provide him or her with "air cover." The executive could then propose a path forward based on the recommendation of an "independent" professional—and may defend their proposal by arguing, "The professionals are the experts; therefore, their findings and recommendations should be respected."

Another example where independence may be compromised is a professional who is an organization's employee and has been given a special (temporary) assignment to perform services. The employee may wonder if their previous position will still be available when they complete the special assignment. He or she may reason that avoiding unpopular recommendations will keep them from "burning any bridges" and increase the probability that they will be able to return to their previous position without worry. This is also an example, expressed in the previous section, of the challenges associated with performing services for an organization as an internal employee.

Temporary Assignment

An example of the second kind of an independence issue is when a service professional has been reassigned to a special project and relieved of their day-to-day responsibilities of their "day job." In this scenario firm managers may desire this second kind of independence (from regular job responsibilities), to allow the professional time to devote his or her full faculties to completing the special project. The direct supervisor may determine that they do not want the professional to spend time or energy focusing on anything other than the special project. Such an arrangement provides a degree of independence, in that it allows the professional to focus on the special project without distraction.

INTERNAL VERSUS EXTERNAL BRILLIANCE

Another phenomenon we have witnessed is the perception that brilliance can only come from high places. This is sometimes referred to as the "not invented here" syndrome. Following is an example. Frank Smith is a former elementary school teacher. During his tenure, he, together with his principal, launched a new science curriculum in their school. The program was based on the notion that teaching science traditionally (e.g., facts from a textbook) was outmoded, especially when considering that the majority of the facts being taught would become obsolete by the time the children graduated from high school. Instead they wanted to teach the process of *doing* science—teach them to do what scientists do. The school became alive with all kinds of science activity. Suddenly, a number of teachers who were uncomfortable teaching science became as excited as the students. They no longer needed to know all the answers. Instead they needed to know the *process*. As a result, the author and his principal were asked to serve on the region's committee to select a new science curriculum. When a committee member proposed looking at this process-based approach, the principal was asked where the team could go see it in action. The answer was, "Come to our school." The response was, "No, I mean, where can we go to *really* see it?" The principal responded, "At our school!" The committee member said, "I didn't know this was going on in your school! I never *read* anything about it." The notion that a revolutionary concept could have organic origins was totally alien to him. In fact, teaching science as a process did not originate with Frank. He merely implemented it. Had it not been implemented "under the radar," management probably would have insisted that the idea be vetted first by a university professor. After all, how could such a radical notion have originated within the rank and file?

It is our belief that some decisions to not use internal resources reflect this bias: "If this guy is so great, why is he working here? Why isn't he on the outside charging big bucks?"

UNUSUAL OR UNEXPECTED INFLUENCES

Despite pre-proposal risk assessments, you might run into unusual or unexpected influences from the client and/or the subject of your work once you are selected for an engagement. Situations might include the following:

- *Media*: If the client has received media coverage, particularly due to problems in the organization, the local media might start asking for interviews and early release of findings and recommendations. It is important not to take the bait but to discuss only the scope of work, project objectives, and process until you are ready to present defensible findings, alternatives, and recommendations.

- *Unexpected Stakeholders*: Particularly for public-sector work, the organization or topic being reviewed might have unexpected stakeholders, such as adjacent communities, neighborhood associations, area businesses, or agencies served by the client. These parties bring their own interests and objectives to the project, which may differ from the client's.
- *Management Changes*: Between the proposal and engagement start-up (or midengagement), we've seen management changes at the upper or mid-level that present new hopes or new problems for the organization. Much of the new manager's approach, philosophy, and ideas may be unknown and untested, particularly if they are from the outside.
- *Last-Minute Blowups*: The scope of work could expand or be redirected if something important (often bad) happens between the time you submit your proposal and project start-up. For example, one of our authors had an engagement with a large city government, which was affected by two visible, negative events that occurred in two city departments during the study. The city then asked the firm to expand its scope to include findings, alternatives, and recommendations to prevent such events from happening in the future.

Unstated Agendas

Although initial discussions or a request for a proposal may identify the scope of work, as well as project issues and objectives, some important actors may have held back or not participated in the definition of the work to be performed. Their agenda may exceed or contradict what has been formally presented by the organization.

For example, two of the authors worked on a project for a library's board of directors, which seemed to focus on the nuts and bolts of board policies, training, and budgeting. But once the preliminary report was delivered, it became evident that the real agenda was to empower the board, relative to the director, with hopes that the director would resign—which she did soon after. At that point, the board thanked us for our good work and terminated the engagement, as they had accomplished what they *really* wanted.

Confusion about Scope and Objectives

Initial discussions about the professional service or a request for a proposal may be skimpy in terms of details, scope, and issues. Despite a solid effort at a definitive proposal, the client may seek to expand, contract, or diverge from your proposal. It is important to hold the line against scope creep, unless that creep results in additional fees and expenses.

The service professional must be able to distinguish between the client who is deliberately vague about the scope of a project (so they can engage in intentional scope creep without paying for the additional services); the client who is confused or uncertain about their needs; and the client with whom you have built a trusting relationship in the past and can reformulate the scope to create a win-win situation. In the first case, you have to draw a hard line in the sand and immediately resolve the scope creep problem. In the second case, you have to invest in educating the client about their needs, through the proposal and during the meeting to review your proposal, in hopes the client sees what they really need. In the third case, you have an established relationship with the client; they trust you. You then will engage in an active discussion with them about what they want you to do; you document it in an engagement letter that is a single-page document laying out the revised scope of work in broad terms; and they sign it. Always bend over backward to preserve the trust you have established with those clients.

COMPRESSED SCHEDULES

Some clients can't wait for your recommendations. A project typically taking four months may be requested in two months. In the public sector, a four-month procurement cycle may allow only two months for the work to be completed. *Beware!* A short project schedule can mean the client doesn't know what it needs; the client is an unsophisticated buyer of professional services; and/or another firm is preselected for the work based on a methodology that you might determine unsuitable to meet your professional standards.

For example, a local government in Illinois asked that a major review of a department be completed in two months. The scope of work clearly required more staff time than that. Later, it was learned the government selected a local accounting firm with little experience, but with close ties to government officials. For that firm, the project was an opportunity to put unassigned staff to work as a loss leader, to add to their potential list of references.

In another example, a chief information officer of a regional supermarket chain wanted to skip the process of defining requirements as the basis for evaluating competing software solutions to manage the operations of his company. He said, "You've done this countless times. You know all of these products. Just save us the time and money and tell us which one is the best." The service professional was able to convince the client that it was in their best interest to follow a process that encouraged stakeholders in the company to own the decision process and the selection of the software solution. Otherwise, it would be management's or the service professional's new computer system, and not that of the stakeholders—who, therefore, may not have a vested interest in implementing the new computer system. Lack of commitment on the part of the stakeholders is at the heart of many system implementation failures.

In any event, service professionals must have a firm idea of their appetite for assuming risk, and then structure the engagement within the acceptable risk limits. It pays to take the time to explain to the client why they need to be fully engaged in the defined process, as well as the value to be derived (or the risk to be averted) by doing so.

FACILITATING CHANGE

Cultural accommodation is the cornerstone of a change in process. The first priority is to develop trusting relationships with everyone affected by the engagement, including the most junior people in the client's organization. If an employee perceives that management (and, by extension, the professional service firm) really doesn't care what they think, then it is very likely that the employee will return the same amount of loyalty.

We have learned that one of the ways a professional can build trusting relationships is to display a style that seeks to foster collaboration and empower individuals to perform. This style has been described as a professional acting like a coach. A management style that employs effective coaching requires professionals to hone skills that may seem alien, when they are used to a command and control management style. A coaching management style that is rich in consensus (not necessarily in unanimity) and empowerment results in employees who excel in both commitment and competence. Many service professionals are rich with experience describing effective coaching skills, such as follows:

- Active listening (shutting off your own thoughts and being in the moment)
- Positive communication, using concrete language (being specific)
- Using descriptive language (focusing on behavior, not personal characteristics)
- Assuming the other person's perspective (how they view the situation)
- Direct perception checking (offering several options for observed behavior)
- Paraphrasing what the speaker said (giving feedback)
- Questioning (open ended to get more detail)

It has been our experience that this *coaching* management style has a higher probability of achieving optimal results. Put more simply: The more heads that are in the game, the better the outcome.

THE SHIFTING LANDSCAPE

All kinds of obstacles appear in the path of engagements. The ability to accommodate and adjust to a shifting landscape may require the professional to be flexible and creative in their approach to getting the engagement completed successfully. In fact, the rapidly changing scenarios in today's fast-paced professional service delivery landscape require firms to establish a set of tools and techniques specifically designed to accommodate rapid change by supporting the service delivery professional.

NEW PARADIGM: RAPID CHANGE AND RAPID DEPLOYMENT

Rapid *change* may be defined as one or more changes that may take place in a service delivery environment where (1) the changes occur in a time frame shorter than the amount of time allowed for in the professional service delivery proposal or contract, and (2) the changes require the scope of the engagement to be redefined. Rapid *deployment*, on the other hand, may be defined as the ability to alter the definition of the service to be delivered and/or the time frame for delivery, to accommodate a change that has occurred in the client's operating environment. Service professionals need to incorporate the ability to adapt to the realities of a rapidly changing environment, as well as the role of the service professional within it. Responding to changes of this magnitude requires firms to have a defined methodology and the system of internal discipline and control firmly in place. This is necessary for a firm to be successful and rise above the competition.

Subject matter for many disciplines often is presented in a form that assumes a perfect world; that all things happen linearly the way they are explained in "the book." Engagement management is no exception. It has been said, "A military battle plan is only good until the first shot is fired. Then all hell breaks loose." Engagement management is not unlike a military battle. Engagement plans can be crafted meticulously with neatly defined work breakdown structures for multiple phases, activities, and tasks. However, business does not stand still while an engagement is being executed. All kinds of discrepant events may occur during the life of an engagement (e.g., CEOs are replaced, budgets get cut, a competitor's sudden new product introduction distracts attention, a new division is acquired, a division is sold, personnel leave or get promoted). Such events will occur in one or more forms, and the service delivery professional needs to have a toolset that is flexible to respond appropriately and with enough rigor and discipline to accommodate changes, without sacrificing quality.

We all have witnessed challenges related to implementation of the new federal Affordable Care Act (ACA), which called for the creation of healthcare exchanges that could be set up by each state. Insurance companies were invited to offer their products on these exchanges. The timeline to become operational was extremely tight, which required the insurance companies to modify their information technology development process. The problem was that the states had not yet defined the detailed specifications that the insurance companies needed to develop their applications. The insurance companies had to start developing portions of the application that *did* have defined specifications. They also had to have a plan to incorporate additional specifications, as they became known. Finally, they had to accommodate changes to some of the code that had already been developed. It was appreciated that each new requirement could potentially alter the design of work already completed, thus requiring rework.

Rapid change requires a nimble, agile ability to change course quickly and have the new direction incorporated into the engagement plan. Being agile also requires the professional to accommodate these changes and report on progress at the next status meeting (e.g., the next working day). Rapid deployment may require an iterative approach to engagement management (i.e., doing a quick implementation, trying it out, and then modifying it once issues are identified).

Our experience has shown that sometimes events unfold that require or result in a major change in direction. A high-priority project last week suddenly becomes relatively unimportant, due to a change in the environment. This is where the risk mitigation strategy and plan come into play.

Sometimes the change in environment should have been expected, but management chose to ignore it or convinced itself that it wasn't an issue because of "groupthink" (that phenomenon where insulated members of management convince each other of facts that are not in evidence). Sometimes groupthink is the result of an autocratic management style, where dissenting opinion is punished. Whatever the reason, the outcome may have a major impact on the engagement.

It is a fact that there is a great deal of uncertainty in today's business climate. Management tends to hedge its bets in this kind of environment. Hedging can sometimes take the form of attempts to accelerate projects, for fear that they may be canceled or because there is a sudden change in direction.

Sometimes the underlying motivation for rapid deployment may cross the line into territory that is not healthy for the client. An example of this is a client manager who elects to fast track a project so much that internal disciplines and controls are bypassed altogether, for fear that competing demands on limited resources may result in the project being canceled or shelved. We would characterize such a scenario as very high risk. In the end, it is up to the service professionals to assess their own appetite for signing on to an engagement with such a high-risk profile.

The notion of rapid deployment has taken on greater significance in recent times, in response to comparisons between implementation life cycles and product life cycles or the life cycle of a selected technology. We have witnessed situations where companies have completed a multi-year implementation only to receive notification from its software vendor a year later that it is discontinuing support for the current version and plans to migrate to a new technology platform.

In the early days of Material Requirements Planning (MRP) systems, users were classified by independent assessors as either Class A, Class B, or Class C. Class A users were judged the best at using the MRP systems to plan, build, and deliver product to customers on time, consistently. A study showed that the majority of these Class A–rated companies kept their implementations simple. They avoided all the bells and whistles and stuck to the basics, while management focused attention on doing the basics correctly. They reasoned that "bells and whistles" distracted attention from the basics. Equally important is the fact that the companies that were judged to be Class A users completed their implementations quickly. Classes B and C, obviously, fell into the trap of designing the perfect (and complex) system, rather than the simple (and very functional) one. The result often is failed implementations or difficult-to-use solutions.

TIPS FOR SOLE PROPRIETORS AND SMALL FIRMS

Smaller firms have an advantage in the new paradigm. Being small allows you to be nimble. Larger firms may not have that same level of flexibility. Here are some tips for dealing with the "real world":

1. No two engagements are the same, even if you are doing the same thing for a very similar client. Make sure you tune in to every client.
2. Develop a clear perspective as to whether you are positioned as an internal or external professional. An example of an external professional being positioned as an internal professional appeared in one of Steve's nine projects for a large Midwestern city. Although he was an independent external professional, he assumed a role that made him appear as an internal professional, in the sense that he was on site each week, had his own desk and phone number, and attended meetings like he was an employee. But he was

an external consultant and expected to be both independent and objective, advocating for the department while also looking out for city interests.

3. If you are an internal professional (e.g., internal auditor, policy analyst, or strategic planner), find ways to protect your independence and objectivity. For example, act in a formal manner when interacting with other departments and staff on projects; have a formal methodology for conducting the project, just like an external professional service provider; and bill your "client" for the work you perform.

4. If you are an external professional, avoid thinking of yourself as an hourly worker, even though you may bill by the hour. Instead, view yourself as adding significant value to your client, aside from the fact that your rate needs to support firm overhead, business development, and internal projects that can't be recovered through direct billing to client work.

5. Commit to being flexible and adaptable, even if it feels counterintuitive. Practice bending on small things, without violating your professional standards. For example, delegate some part of the project to a subordinate, particularly one of your favorite tasks; or say yes to a client request that might seem a bit unreasonable, while being mindful of number 6 below.

6. Resist "scope creep" or take advantage of it by asking for a larger role and budget.

WRAP-UP

Adapting plans to accommodate real-world events and situations is something service delivery professionals must be able to do to meet or exceed client expectations. It begins with having a firm view of who the client is and who their customers are.

Next, you need to have a clear understanding of your position within the client's organization; equally important is that the client's perception and understanding of your position needs to be managed. Further, you must be cognizant of the nuances that may affect a change in the client's perception of that position during the course of your engagement.

Another factor for consideration has to do with accommodating a client's desire to fast track an engagement or sudden changes that impact the delivery timeline or the scope of work to be done.

Finally, each of these real-world factors needs to be accommodated, while not compromising adherence to those elements that define internal discipline and control or your firm's professional standards. Saying no may be in the best interest of your firm, even if you lose a client or an engagement.

6 Conducting the Engagement

Rules for Conducting the Engagement

Rule 2: Develop and *implement* a system and culture of internal discipline and control, to ensure consistency of service, efficiency of operation, and quality and reliability of deliverables. Then train, mentor, and monitor personnel regarding engagement management policies and procedures.

Rule 3: Establish and enforce engagement documentation standards, including those for proposals, progress reports, and deliverables.

Rule 6: Be flexible and adaptive to the "real world" once the project starts, to manage the dynamic between client expectations and what's *really* happening within both the firm and the client.

Rule 7: Implement a provider–client communication plan that will ensure clear and frequent discussion of engagement progress and status.

Rule 8: Bill and collect frequently. This will improve cash flow and alert you sooner if the engagement is in trouble with the client.

Rule 9: Conduct independent status/quality reviews of the engagement *while it is in process* and subsequent to conclusion. Involve key client contacts in the reviews.

Now the fun part of the engagement starts! You've sold the engagement. In the proposal you've laid out the approach, work plan, schedule, and budget. You've developed a risk mitigation plan, as well as a communication plan—and folded it into the work plan and schedule. And you've selected and prepared the engagement team to conduct the work.

But doing the technical piece of the work is the easy part; after all, that's what the firm is all about. Managing the engagement so the firm makes a fair return on the work, while maintaining a fully satisfied client—*that's* the difficult part. But if you have (1) a well-thought-out work plan and schedule that considers key risks, (2) a realistic engagement budget that reflects what the client is willing to pay, (3) a communication plan that keeps both the client and the firm apprised of progress, and (4) a strong culture of internal discipline and control (IDC), then management of the engagement is simplified.

The remainder of this chapter deals with the ways in which the engagement can be managed just before it begins and while it is in process. It does not cover the technical aspects of the work itself, which will *vary considerably from one professional service firm to another.*

PRE-KICKOFF ACTIVITIES

With the engagement awarded and contract signed, a number of tasks must be completed *before* you go on site to start the engagement.

First, you should conduct a preliminary telephone interview with key client representatives to confirm the scope, project activities, and project schedule, particularly if there is a gap between the proposal and award dates, as often happens with a government client. This also can be an opportunity to better understand the client and its needs, as well as to fine-tune the work plan, risk mitigation and communications plans, and project budget allocations to people and tasks. Plus it opens up relationships with client staff whom you may not have met during the proposal preparation and presentation process. In most cases, decision-making executives select the professional service firm (you probably met them when selling the engagement), but upper to midlevel management primarily are involved as the engagement unfolds (you meet these folks once the project starts—and the sooner the better!).

Second, prepare a handout to review the communication plan with the client at the kickoff meeting.

Third, update your standard information request to match the requirements of the engagement. If possible, collect key documents in advance of the kickoff meeting, so you arrive on site knowledgeable of key activities, resources, and issues relating to the client. For example, information requested for a management consulting engagement is listed below; but note that this list will differ across the various professional service sectors and for specific engagement situations:

- Legal documents (incorporation, charters and codes, contracts and agreements)
- Strategic, financial, operating, technology, and other plans
- Budgets and financial reports for the past two or three years
- Organization chart, staffing plan, and job descriptions
- Compensation plan, personnel policies, union agreements, and performance evaluation system
- Operating policies and procedures, workflows, reports, and statistics
- Inventory of facilities and equipment
- Inventory of technology applications and equipment
- Studies and analyses related to the project

Fourth, schedule initial meetings, including the kickoff meeting, and set up on site logistical arrangements for the project team (access passes to client offices, work space, computer use, Internet connection, copier access, and the like).

ENGAGEMENT KICKOFF MEETING

In Chapter 4, Setting Up the Engagement, we described the importance of the initial meeting with the client to start the engagement. We provide additional details in this chapter.

The primary purpose of the engagement kickoff meeting is to continue to build a solid relationship between the firm's engagement team leaders and key client contacts (i.e., those people who are major stakeholders in the engagement outcomes). Of course, relationship building should have begun during the proposal process, the presentation to review the proposal, and any other meetings or communications leading up to the acceptance of the proposal (as described in Chapters 3 and 4). The development and maintenance of sound, high-quality relationships based on trust are important in any professional service endeavor.

The first question is, Who should attend the kickoff meeting? Typically the firm's representatives should include at least the engagement director and the project manager, as well as team leaders in larger projects. The firm's client relationship leader also may wish to attend, particularly if he or she knows any of the client's attendees. From the client's side, those people who have a major stake in the engagement's successful completion should be invited to the meeting, including the members of the client's engagement steering committee (if established yet), the project liaison (your go-to person for logistical help), and key managers and stakeholders.

The topics covered in the kickoff meeting should center on the objectives of the engagement, the related deliverables, the approach to be used, the timetable, and potential risks. Following is a typical agenda for the kickoff meeting:

- Introduction of the participants, including their roles in the engagement and their positions in the firm or the client's organization
- Discussion and confirmation of the engagement's objectives and deliverables (the client's expectations for the firm)
- Explanation of the firm's approach and work plan to be employed, which might need to be adjusted now that the engagement is under way
- Review of the schedule for the overall engagement, as well as any phases involved
- Review of the communications plan, including methods to communicate progress reports and deliverables (electronically is becoming more common)
- Discussion of any potential risks, issues, pitfalls, or obstacles in conducting the engagement, such as "unhappy campers," inexperienced managers or employees, pockets of resistance to change, past situations on similar projects, union matters (if any), and financial limitations to investments that may be needed as a result of the engagement
- Resolution of any questions that may be on anyone's mind

If not already in place, a final item on the agenda should be the formation of an engagement steering committee composed of both firm and client representatives.

Such a committee would be charged with monitoring the engagement as it progresses and moving it along when barriers are encountered or changes are necessary. The steering committee's members typically would be engagement supporters, although antagonists may also be included to ensure that their views are heard and, to the extent possible, accommodated. In some circumstances, particularly in the public and nonprofit sectors, the steering committee may include elected officials, board members, and citizen/customer representatives.

The engagement steering committee could have a number of roles. For example, it could provide direction to the engagement process and team; review and approve emerging issues and formal deliverables; receive periodic reports regarding engagement progress, status, difficulties, and near-future activities; help clear roadblocks as they occur; review draft deliverables; and provide final approval that the engagement has been completed satisfactorily.

Some professional service engagements may require additional committees to review very specific or technical deliverables, such as follows:

- Audit reports
- Engineering or architectural designs
- Legal positions or responses to lawsuits
- Marketing plans with draft brochures, website updates, and target audiences
- Position reclassifications or reassignments
- Reengineered work processes
- Information technology needs and requirements
- Equipment and facility changes

The engagement steering committee may not have the detailed knowledge to comment effectively on these detailed topics. These technical committees could include staff from finance, human resources, information technology, legal, marketing, operations, and other functional departments.

In association with the kickoff meeting, the engagement director, project manager, and team leaders also may want to schedule informational briefings on the scope and process of the engagement with client managers in the areas of finance, human resources, information technology, or other central support functions.

In some assignments, the engagement leaders also may provide a briefing to affected staff on the purpose, objectives, and process for the engagement, particularly if employee input will be sought through data collection, interviews, development of process maps, needs assessments, and the like. These visits could include training on data collection tools, if applicable.

ENGAGEMENT MANAGEMENT AND SUPERVISION

Chapter 4, "Setting Up the Engagement," describes roles of the various members of the engagement team. It is important to ensure that all members of the team are clear on their roles, budget, and schedule for the engagement. Typically, these roles are discussed in meetings in the firm's office before the team moves on site to serve the client.

ROLES OF THE ENGAGEMENT MANAGEMENT TEAM

The primary role of the engagement management team is to hold the overall engagement team together—to keep them productive (both efficient and effective), on schedule, and on budget. Specific assignments for the engagement management team when the engagement is underway include the following:

- *Relationship leader*: Monitors overall client satisfaction, helps solve engagement difficulties, and identifies opportunities for additional client service.
- *Engagement director*: Reviews and approves process and deliverable quality, adherence to the firm's system of IDC, budget compliance, the effectiveness of the project manager and the entire engagement team, contract change orders, and client billings *before they are issued*.
- *Project manager*: Manages day-to-day activities of the entire engagement team as they carry out the work plan, controls the budget for the entire engagement, prepares progress reports, drafts contract change orders, compiles draft and final deliverables, and prepares and issues progress and final bills. These assignments are reviewed and approved by the engagement director, who supervises the work of the project manager.
- *Team leader*: Manages day-to-day activities of team members and tasks assigned to the team, controls the budget for these tasks (under the direction of the project manager), conducts fact finding and analyses related to assigned areas and tasks, and prepares draft task deliverables/reports. These assignments are reviewed and approved by the project manager, who directs the work of team leaders.

The engagement budget must allocate adequate hours and expenses for management and supervision, particularly for the engagement director and project manager. Depending on the nature of the client relationship, the relationship leader also may have hours and expenses assigned to attend key meetings and presentations, discuss progress with the engagement team, and review deliverables.

MULTI-PROJECT MANAGEMENT

It is common for the engagement director, project manager, team leaders, and technical staff, even in smaller firms, to participate on multiple, concurrent engagements involving several clients. It is critical to the firm's success to set up a management structure (who's in charge?) and staffing plan (who does the work?) to deal with current projects. In addition, the firm should estimate minimum and maximum resource requirements over the course of the year to ensure that the firm has adequate staff and other resources to successfully meet its client obligations. Too few resources may cause client work to suffer; too many resources may cause profitability to suffer.

Based on the typical number of engagements and the size of engagements, the firm will need to staff up on project managers, team leaders, and technical staff (or subcontractors) to meet workload requirements. Smaller firms, or firms with less predictable workloads, are more likely to use contract employees or partner firms to meet the ebb and flow of the workload. Larger firms, or firms with more predicable workloads (like audit firms with multiyear contracts or law firms with significant retainer work), are more likely to hire employees.

A key task for each professional service firm is to determine how many active engagements can be effectively managed at one time with current and expandable subcontractor resources to ensure high-quality deliverables and client satisfaction. For Steve Egan, this means an objective of somewhere between three and eight projects per year. A year with a smaller number of projects means larger projects (with budgets over $50,000 and with several consultants), and a year with a larger number means smaller projects (with budgets from $10,000 to $25,000, many of which he might do himself). He tries to limit open projects to three, all in different stages (one just starting, one in mid to late fact finding, and one in analysis/report preparation). Often this doesn't work out so well, with two or more projects running side-by-side—meaning that travel and the workload are crazy. But, this is better than being "on the beach" without chargeable hours.

Frank Smith typically is involved in much larger information technology projects that require his full-time attention as part of a large team for months and perhaps for as long as a year. Therefore he will work on only one or two projects per year. Barry Mundt recently has conducted many smaller projects scattered across the year, but had a schedule like Frank's earlier in his consulting career.

As projects increase in scope and duration, the role of the team leaders may expand so that they become quasi-project managers for engagement segments or smaller projects. This emphasizes the need to train and mentor team leaders as if they were project managers.

MANAGING SUBCONTRACTORS

If you use subcontractors, develop and apply a formal subcontractor agreement that defines expectations, work plan task assignments, hours and expense budgets, on-site time and the number of trips, and deliverables. Subcontractors, just like firm employees, need to be schooled on the firm's IDC policies and procedures. It is particularly important to define the role of subcontractors in briefing clients on emerging issues and preliminary ideas; this will help to ensure what they say meets firm policy and presentation lines and is consistent with the engagement leaders' preliminary ideas for the scope and content in the eventual deliverable. The project manager should be aware of and approve in advance any communications from subcontractor to client. An easy way to mess up a project is to give a subcontractor too much leeway, so that they become something of an independent actor in the eyes of the client. Remember, it's your engagement—not the subcontractor's!

ENGAGEMENT CONTROL PROCESS

Chapter 5 also defines the engagement control process (engagement control system, engagement file, engagement guide, work plan and schedule, budgets and resource loading, risk mitigation strategy and plan, and communications strategy and plan). It's a good idea to review your firm's engagement control process just before each engagement starts, and then tailor it to the specific project, as needed.

MANAGING THE WORK PLAN, SCHEDULE, AND BUDGET

The project manager, in association with the engagement director, is responsible for managing the work plan, schedule, and budget. Most firms will develop a project monitoring and reporting system, which could be a simple spreadsheet or a more complex computer application. Key elements of the monitoring and reporting system are as follows:

- Develop a list of tasks and subtasks
- Enter budgeted hours, billing rate, and total fees and expenses for each member of the engagement team and for the engagement in total
- Enter budget adjustments to hours, billing rate, fees, and expense budget for each member of the engagement team and for the engagement in total
- Enter actual time, fees, and expenses for each task/subtask and for each client bill, along with a budget balance for each team member and for the engagement in total
- Calculate totals for the various data elements
- Analyze budgetary information as needed, which could include the number of trips taken and remaining, the percentage of budget used, and the percentage of total budget billed and collected
- Enter the total contract amount, which may be different than the budget, and any reserve for contingency. In Figure 6.1 below, the budget is $49,000, but the contract is $55,000, leaving a $6,000 reserve for extra trip expenses or overage on hours changed to the engagement.

A sample engagement budget and billing worksheet for a two-person engagement is shown in Figure 6.1.

Sample Budget and Billing Worksheet (Simplest Updated 2-17-2014)

Staff	Categories	Budget	Bill #1	Bill #2	Billed To Date	Budget Balance
Eng Director	Hours	80	28	12	40	40
Assigned two	Rate	$ 150	$ 150 $	150	$ 150	$ 150
departments	Fee	$ 12,000	$ 4,200 $	1,800	$ 6,000	$ 6,000
	Expenses	$ 5,000	$ 1,000 $	1,000	$ 2,000	$ 3,000
	Total	$ 17,000	$ 5,200 $	2,800	$ 8,000	$ 9,000
Project Manager	Hours	200	44	60	104	96
Assigned six	Rate	$ 130	$ 130 $	130	$ 130	$ 130
departments	Fee	$ 26,000	$ 5,720 $	7,800	$ 13,520	$ 12,480
	Expenses	$ 6,000	$ 1,000 $	2,000	$ 3,000	$ 3,000
	Total	$ 32,000	$ 6,720 $	9,800	$ 16,520	$ 15,480
Eng Total	Hours	280	72	72	144	136
	Average Hourly Rate	$ 135	$ 135 $	135	$ 135	$ 135
	Fees	$ 38,000	$ 9,920 $	9,600	$ 19,520	$ 18,480
	Expenses	$ 11,000	$ 2,000 $	3,000	$ 5,000	$ 6,000
	Budget Total	$ 49,000	$ 11,920 $	12,600	$ 24,520	$ 24,480
	Reserve	$ 6,000				$ 6,000
	Contract Total	$ 55,000				$ 30,480
	Billed to Date		$ 11,920 $	24,520	$ 24,520	
	Percent To Date		21.7%	66.3%	66.3%	33.7%

FIGURE 6.1   Sample budget and billing worksheet.

REALIZATION VERSUS UTILIZATION

If your firm bills on an hourly basis and uses billing rates, you will be engaged in the *realization* versus *utilization* challenge. You will have to address this key dilemma on a recurring basis: How do we balance realization versus utilization, both for the firm and for individual employees and subcontractors?

Realization is the percentage of the standard billing rate that you actually bill to an engagement. For example, if an engagement with one team member has a budget of 100 hours at a $250 per hour standard billing rate, the budget for fees is $25,000. Realization then varies depending on the number of hours actually charged to the engagement:

- 67 percent realization of the standard billing rate if he/she bills 150 hours
- 100 percent realization of the standard billing rate if he/she bills 100 hours
- 133 percent realization of the standard billing rate if he/she bills 75 hours

To get work or because of engagement challenges or competition, however, a firm may decide to adjust standard billing rates up or down from one engagement to another. And the standard billing rate might be adjusted annually based on a person's performance or promotion. Steve Egan's rate, for example, floats by about $25 per hour, depending on the size of a given client, the degree of difficulty of the work, the level of risk to the firm (really to him!), and/or other intangibles (such as his desperation for work). So your starting budget may result in a reduced or enhanced realization of the standard rate. Litigation support work, for example, may cause a firm to bill at 150 percent of standard rates, due to the high level of risk involved. On the other hand, a strategically beneficial engagement might be billed at 75 percent of standard rates, because the experience and the reference could be useful in gaining future work at standard rates, or because it keeps people busy for months at a time.

Utilization is the percentage of paid time someone bills. So, if you are paid for forty hours in a week and bill forty hours, your utilization rate is 100 percent. If you only bill thirty hours, then your utilization rate is 75 percent; and if you bill sixty hours, your utilization is 150 percent. But utilization is a tricky thing, because someone on a client engagement may bill at a fully utilized rate, while someone on a mission critical, nonclient assignment for the firm (such as development of a new service area) may not be generating any current revenue but be fully engaged in the development work. Both, however, are doing important work that benefits the firm, either right now in client revenues or in the long term in new service area revenues.

If the firm has its focus on standard hourly rate realization, staff may undercharge hours to keep their realized billing rate high; however, in that case, revenues will decline. If the firm focuses on utilization, staff may overcharge hours to keep their utilization rate high, but the budget will get used up too fast (if there is a fixed-price contract). The key to meeting this challenge is to stop focusing *either* on realization *or* on utilization and consider both factors in judging staff productivity and profitability at your firm.

CHANGE ORDERS OR CONTRACT AMENDMENTS

The scope of an engagement may change after it is under way, thus requiring a change order or contract amendment. The project manager, therefore, should watch the work being performed to identify "scope creep." At some trigger point (defined by each firm), the engagement director and project manager should contact the client to discuss a formal change in scope, schedule, and budget. These change orders should be documented and approved in writing by the client.

When to ask for a change order or contract amendment is an art form that each firm needs to consider in developing its IDC system. You don't want to upset the client with constant requests for small changes to the contract, but you also don't want to accept a bunch of small-scope additions from the client that mess up profitability or your ability to serve multiple clients. Basically, it's a value judgment. One approach is to ask for a contract amendment if the client requests (or sneaks in) a large task or several smaller tasks that cause, say, a 10 percent increase in the budget. That approach implies that the firm is willing to do what they said they would do for the client—within reason. Any such change will require a change in the work plan, schedule, and budget, with appropriate approval by firm and engagement leaders, as well as potential updates to subcontractor agreements.

MANAGING THE ENGAGEMENT FILE

The firm needs to define some structured way of organizing physical and electronic "stuff" collected and generated during the engagement. Having moved from a very large, international audit, tax, and consulting firm to smaller practices, the authors have continued to apply a consistent engagement file strategy that involves these key elements:

- *Designate a keeper of the records for each engagement, most likely the project manager.* Make sure that this person is fully capable of complying with the firm's engagement file policies and procedures.
- *Define the role of each team member, particularly team leaders, in adding to and maintaining the master record.* This task is particularly important if team members are in and out of client offices at different times and/or when the team is from multiple offices.
- *Structure the engagement file to match the scope of the engagement.* A smaller project might need only a few file folders to maintain the records; a larger one may need a cataloging system to keep track of boxes of information. In either case, use the same basic structure and sequence of records for every project.
- *Keep the engagement file up-to-date* with at least weekly reviews of all information in the file and information recently collected to ensure the file is complete.
- *Define ways to catalog and manage electronic records*, such as client budgets and financial reports in the initial information request.
- *Consider the value of an engagement website or file-sharing system for internal use only.* For example, Dropbox is what we have used to compile this book; Frank Smith regularly uses Dropbox to manage complex information technology projects.
- *Consider the value of an electronic input process for client staff* and define how it will be managed. For example, one could set up an e-mail address specifically to gather client comments on an engagement.

- *Figure out what to keep in the engagement file and what to discard after the engagement has been completed.* For example, sensitive interview notes and employee surveys may best be discarded, particularly if the professional service firm has made a strong promise of confidentiality. Note that the engagement file of one of our authors' projects was subpoenaed due to a management–labor dispute. When the lawyers call, it's too late to clean things up and throw things out!

FIRM PROGRESS REPORTS

The firm should prepare periodic engagement progress reports that particularly relate to quality assurance and risk assessment. Internal firm progress reports will differ from those that are prepared for the client.

The relationship leader and engagement director will expect reports from the project manager and team leaders on a regular basis. The project manager should informally caucus engagement team members at least weekly (perhaps more often in a large engagement) to catch up on their work and, most importantly, to confirm they are on schedule and budget. The project manager then should prepare at least a simple report on the engagement's status for the relationship leader and the engagement director. Key questions to be answered in the internal progress reports are listed below:

- Are we on schedule and budget?
- Are senior client personnel satisfied with our work?
- How well is the engagement team working together? Are there any personnel issues to resolve?
- Are there any personnel issues to resolve?
- Are there any bumps in the road so far? If so, how will they be resolved?
- Are there any emerging issues arising, particularly any that are not "routine"?

Frank Smith sometimes uses a formal project management methodology to address the first question. While this methodology was first used with information technology development projects, its use has been expanded into other disciplines. A key benefit of this approach is that it allows for near-continuous updates regarding budget performance, with the "budget" typically expressed in hours. Given the anticipated hours associated with each identified task over a period of, say, thirty days, the number of hours remaining will give a quick, daily status update on budget versus actual hours, as well as schedule compliance. It is simply a comparison of the number of budgeted hours (based on the number of tasks to be completed at a given milestone) versus the actual hours expended for tasks to be completed at that milestone. A casual observer can determine if the project is ahead of, on, or behind schedule. A project status meeting is conducted on a daily basis for about fifteen minutes at the beginning or end of the workday. For Frank, these reviews are referred to as *scrum sessions* and are led by a *scrum master*.

CLIENT PROGRESS REPORTS

Similarly, clients expect progress reports from the engagement leaders to ensure the engagement is on time and on budget and it is meeting their objectives and expectations. A typical client progress report will include the following:

- Discussion of work accomplished in this reporting period
- Work to be accomplished in the next reporting period
- Identification of any issues, delays, or obstacles to meeting objectives and staying on budget and schedule
- Emerging issues and ideas
- Presentation of draft task, segment, or phase deliverables

The size of the client's project will drive the frequency and complexity of these reports. On smaller engagements, the progress report will be short and to the point. Such reports often can be imbedded in a cover letter to the progress bill and presented as a sentence or two on each of the above topics. Issues and delays, however, should be presented in more detail, along with a corrective action plan. On larger engagements, the report will be more complex and likely include charts showing hours and cost expended by task or subtask.

For all engagements, written client progress reports are released in advance of progress meetings with key client staff; at a minimum the report should be submitted to the engagement steering committee. Depending on the nature of the engagement, progress meetings should be held at least monthly (for smaller engagements) and perhaps more often (particularly on larger engagements). Following the progress meeting, the client and the firm should leave the room pleased with the progress of the work and with any corrective action plans deemed necessary to accommodate scope changes or issues/delays.

Progress Bills

Progress bills should be issued at least monthly; larger engagements that consume considerable resources may require weekly or biweekly billings to improve cash flow. The sequence of billing is defined in the proposal/contract and typically follows a set schedule (e.g., 10 percent biweekly) or is based on actual resources expended in a billing period.

Steve Egan's pattern is to bill 10 percent at project kickoff to get established as a vendor in the client's accounts payable system (i.e., to "prime the pump"), followed by additional monthly or milestone bills up to 80 percent or 90 percent of the total project budget. The final 10 percent or 20 percent is billed on delivery of the final report and presentation. His objective is to keep pace with the expenditure of resources to facilitate cash flow, particularly to cover expenses when he and other team members have to pay them.

Quality Assurance and Quality Control Activities during the Engagement

Based on the engagement's risk mitigation plan, the relationship leader, engagement director, and project manager will conduct quality assurance and quality control activities during the engagement process. Quality assurance (QA) focuses on the engagement process, and quality control (QC) on the engagement product or deliverable.

The following QA activities relate to engagement plans and the process of conducting the engagement:

- Review and approve the final work plan, schedule, and budget, including assignments and budget allocations to engagement team members
- Review and approve the final risk mitigation and communications plans
- Monitor hours used and expenses incurred, as reported weekly or at key milestones (such as monthly, completion of work plan tasks, or when the client is billed)
- Review the performance of team members and develop corrective action steps (including the replacement of team members, in the extreme case)
- Review technical work in progress (e.g., detailed designs or subsystem operability in an information technology engagement or the quality and quantity of information gathered from fact-finding interviews and surveys in a management consulting engagement)

In addition, QC activities will focus on review and approval of *all* client deliverables (draft and final), *before* they are released to the client including,

- Emerging issues and ideas
- Progress reports to the client to ensure they comply with the proposal
- Draft client deliverables, including handouts at client meetings
- Final deliverables
- Final presentations to the client

Depending on the scope and complexity of the engagement, as well as the firm's size and volume of work, these QA and QC activities could be relatively informal or very formal. It is important that each professional service firm define the level of diligence required to meet firm and client expectations. Larger engagements that consume more staff time and the work of larger firms typically require a very formal QA/QC process. On the other hand, smaller engagements and the work of smaller firms tend to be less formal, typically requiring only an e-mail or short phone call between the engagement director and project manager for emerging issues and progress reports, as well as a detailed, face-to-face review of handouts, draft deliverables, and final deliverables. In both large and small firms, it is important to document your QA/QC process and train leaders and staff on the requirements of the process.

DELIVERABLE PREPARATION AND PRESENTATION

No matter how great of a job you do in collecting facts and information and then analyzing them, the deliverable becomes the heart of each engagement and the basis for the client's assessment of your work (see Appendix B, Deliverable Preparation Guide). Although the deliverable can be a short letter with accompanying information (e.g., architect's drawing or insurance policy) or a longer document with several sections (e.g., executive summary, technical report, and appendixes), what you produce needs to be accurate in the facts presented, clear in presentation, well written, understandable, compelling, and implementable. And the process used during the engagement—particularly progress reports, emerging issues, and draft deliverables—should build momentum for and commitment both to the final deliverable and to post-engagement actions by the client.

APPROACH TO DELIVERABLE PREPARATION

Although you could wait to the end of fact finding and analysis to prepare your draft deliverable, many professional service firms follow a step-by-step process to develop foundational elements of the ultimate engagement deliverable. These elements could include the following:

- Preliminary analyses of industry trends, operational and financial data (e.g., the growth in budget and personnel over time), employee surveys, and interviews (e.g., analysis of strengths, weaknesses, opportunities, and threats)
- Identification of emerging issues that are presented to the client for reaction, correction, and clarification in progress meetings
- Preliminary or progress deliverables that may be phases, chapters, or specific elements of the ultimate, comprehensive deliverable

Using a step-by-step approach to deliverable preparation has several advantages and is useful on both large and small engagements:

- Gives the client deliverable-related information in small doses to help them absorb it easier
- Gives the engagement team the opportunity to assess the client's reactions to preliminary findings and ideas early in the engagement (particularly to identify and work on naysayers)
- Pulls the team's thoughts together while information and ideas are fresh
- Spreads out the deliverable preparation effort across the engagement, rather than being clustered at the end of the engagement when a lot is going on and the end of the schedule is imminent

DELIVERABLE PREPARATION CHALLENGES

Appendix B, Deliverable Preparation Guide, provides a more extensive guide to developing deliverables. Key practical matters to consider when conducting the engagement and preparing the deliverable are as follows:

- *Who should prepare the various elements of the deliverable?* On smaller engagements, it's best to have one preparer. On larger engagements, several people will provide parts of the deliverable to a single compiler. In the end, one or two people—typically the engagement director and project manager—must compile and fully understand everything in the deliverable to be well prepared for meetings that present and review the report with the client.
- *How can the deliverable reflect "one voice" if there are several preparers?* A Report/Deliverable Preparation Guide should be a part of the IDC system to provide all service professionals within the firm and subcontractors with direction on how to structure and prepare a report/deliverable. For example, a basic primer on grammar (e.g., which versus that, who versus whom), punctuation (e.g., the frequency of commas), and writing style (e.g., text or

bullets) could be prepared for a written report. During the engagement, one person (typically the project manager or the engagement director) must be the final compiler and editor to achieve that one voice.

- *How does the deliverable preparation sequence and schedule fit into the engagement budget and schedule?* In particular, the engagement leadership must define how much time (in team member work hours and in schedule days) is needed for internal review before delivery to the client. And, based on the size and complexity of the deliverable, the engagement leadership and the client should define how much time is needed for the client to review the deliverable before a draft review meeting or final deliverable presentation. Be wary that late deliverables will rile up a client, even if your work so far on the engagement is great and your draft and final deliverables are wonderful.

- *What should be in a written deliverable and what should be left out or presented privately?* A part of this question relates to the length of the deliverable (see the next bullet point). The primary issue, however, is dealing with sensitive information, such as poor performance of individuals, cultural and relational issues, health or safety questions, negative financial results, land purchase decisions, and "doomsday" issues or events (if doomsday events occur, really bad things will happen and the client's future may be threatened). These issues might better be communicated orally in a meeting with senior executives, before they are included in the official deliverable (if at all).

- *What length of deliverable is needed?* Generally, the shorter the better, but the deliverable needs to answer all client questions and provide all relevant data and analyses. Some firms adopt a policy of presenting major ideas in bullet point format, and that's it; others take a more analytical approach, with lengthy written deliverables, perhaps in multiple volumes. You need to decide how your specific professional service should be presented to meet the needs of your firm and your clients.

- *How do you structure the deliverable* (e.g., short letter perhaps with attachments, detailed technical deliverable, or multi-document deliverable)? Appendix B, Deliverable Preparation Guide, provides a guide to the structure and content of an engagement's deliverable.

- *How do you structure the executive summary,* which is included in *every* report in some fashion? In a letter report, the executive summary would be the first section of the letter. On larger reports, the introductory letter can present key points, Chapter 1 can be the executive summary, or the executive summary can be a separate document. The report structure is driven by the scope of work requested, technical issues addressed, and the interests of the client.

- *Who gets the deliverable, both draft and final (which may be different audiences)?* The preliminary draft deliverable usually is shared with a small group (e.g., the engagement steering committee and perhaps a few others). This allows the deliverable to be "fixed" before it is released to a larger audience, which may include the client's governing board, senior management team, and/or employees. In the public sector, that larger audience could include citizens, business leaders, and other stakeholders in a public meeting that's shown on cable TV.

PRESENTATION OF DELIVERABLES

When a deliverable is prepared and submitted to the client, you should hold a meeting to go over it with key client personnel, usually the engagement steering committee. Sometimes the meeting is fairly short, going over major issues, ideas, and supporting documents; in other cases it will be longer, with a page-by-page review of the entire deliverable document. Your first decision involves what approach to take—highlights of major ideas or a detailed review of the entire document. The scope of work and the audience will drive the approach. In most cases, technical managers will want details, and senior management and board members will expect only the highlights, so you may need to schedule two review meetings.

The meeting to review the *draft* deliverable likely will be more technical in nature, with limited participants. The time committed by the client to the review of draft deliverables may inhibit a full review of the document. One approach for a written report, with time permitting, is to ask for general comments or criticisms of the deliverable, followed by a section-by-section or page-by-page review. If time is short, do the general review in the meeting and ask each participant for specific comments, compiled by the client's project liaison, in a week or two.

The meeting to review or present the final deliverable will be more formal than the review of the draft deliverable, and likely will involve senior executives and perhaps the organization's board of directors. In these meetings, the firm usually has a strict time limit (often thirty minutes or less), so a crisp, bullet point presentation of key ideas by the engagement director and project manager is often the most practical approach. For example, the engagement director would review the engagement's purpose and the broader issues, while the project manager would present the details of specific deliverables (e.g., organizational recommendations, facility design, legal documents, or insurance coverage).

After introduction of the meeting participants, you should thank the client for the opportunity to serve them, acknowledge the contributions of client staff, and ask if everyone has the entire deliverable—which helps you judge the audience's grasp of the results of the engagement. The engagement director is in charge of judging audience reaction (understood or confused, agreed or not, engaged or bored) and may jump in to rev things up. At the end, again thank the client for the opportunity to serve them and ask if there are questions. You need to decide, based on the audience and time allowed, if questions can be asked during the presentation or held to the end.

TIPS FOR SOLE PROPRIETORS AND SMALL FIRMS

Smaller firms, including sole proprietors, have unique challenges in conducting engagements, because key leaders often play multiple roles (e.g., relationship leader, engagement director and project manager, and staff) on several concurrent engagements. Key firm objectives are to ensure that the unique responsibilities of these roles are performed well, which can be a bit of a balancing act for a one- or two-person firm or a small engagement team. Typical strategies for small firms in conducting the engagement are listed below:

1. Train all staff, including subcontractors, on your engagement policies, procedures, and rules; then monitor behavior closely to ensure compliance.
2. Invest time off-engagement in setting up recurring engagement needs—the initial informational request, interview guides, report structure, and presentation models—so you don't have to reinvent them again and again during engagements.
3. Define your limits to ensure success, including the number of engagements you can handle at a time, size limits for engagements (particularly what's too big), and the number of staff you can manage effectively. Overreaching is the quickest way to failure.
4. Relentlessly manage the work plan, budget, and schedule to meet client expectations. Set up a recurring schedule (typically weekly) for the engagement director and the project manager to review of these key control processes.
5. Define realization and utilization expectations for the firm and staff members that are realistic and support profitability.
6. Be a hawk on scope creep and the need for change orders or contract amendments. Define hard limits that trigger discussions with the client about budget adjustments. For example, if they ask for an extra trip, make sure you tell them they'll have to pay for it (time and expenses).
7. Keep your engagement files up to date, particularly related to administrative matters such as contract changes, progress and final bills, and project management notes.
8. Prepare internal progress reports even if they are distributed only to yourself and the engagement file's project management folder or an e-mail folder. This practice allows you to look back at the history of the engagement to document key facts, decisions, or issues.
9. Set up times to conduct the level of QA/QC that works for your firm. Perhaps use Saturday mornings to step back from engagements and review how they are going.
10. Whittle down your model engagement deliverable so it is as short as possible, while covering all issues in the scope of work. Prepare recurring sections in advance so they can be inserted, rather than rewritten for each engagement.
11. Bill early and often!

WRAP-UP

We encourage you to see the entire process of delivering your professional service as being the end product. Success on an engagement is determined not only by the final deliverable, but also by

- Personal trust relationships built up during the engagement
- Informal interactions outside of official meetings (e.g., lunches in the client's cafeteria and break room chats)
- The quality of progress reports and meetings

- The quality of draft deliverables for various phases or subtasks in the engagement
- Your ability to use these interim steps to build the client's excitement and commitment to the eventual deliverable

Again, being on time and on budget is critical for any professional service engagement, particularly if the client has expressed in advance a schedule expectation or limitation. Newspaper articles often tell of information technology projects that are years past schedule and millions over budget, resulting in the termination of contracts and a bad reference and reputation for that firm.

7 Closing the Engagement

Rules for Closing the Engagement

Rule 3: Establish and enforce engagement documentation standards, including those for proposals, progress reports, and deliverables.

Rule 8: Bill and collect frequently. This will improve cash flow and alert you sooner if the engagement is in trouble with the client.

Rule 9: Conduct independent status/quality reviews of the engagement while it is in process and *subsequent to conclusion*. Involve key client contacts in the reviews.

Now you're on the two-yard line but need to punch the ball in to score the final, winning touchdown—closing the engagement and moving on to your next client. Some important final tasks must be performed before the engagement is closed out and put to bed:

- Preparing and submitting final deliverables
- Final billing and collection
- Final review of the engagement with the client
- Wrapping up engagement documentation and files
- Conducting an independent evaluation of the conduct of the engagement

All of these tasks are important to the key goals of gaining the client's full respect for the current engagement, securing a solid reference, and developing client interest in future work by your firm.

FINAL DELIVERABLES

Chapter 6, "Conducting the Engagement," provides specific guidance on preparing engagement deliverables, and Appendix B provides a guide to preparing deliverables. The final deliverables for a professional service engagement will be a final letter, which transmits technical deliverables such as engineering designs or architectural plans, or a final report, which is common for a management consulting or other analytical assignment. The final deliverables often include a final presentation

TABLE 7.1
Steps to Ensure Deliverable Acceptance

1. Get the client's formal approval of the final draft and permission to "go final."
2. Get the firm's relationship leader and project director to formally approve going final.
3. Prepare a master copy of the final deliverable.
4. Have an independent party, who is not part of the engagement, review the final deliverable to ensure that it is readable and grammatically correct, spelling is correct, tables and figures add up, and it is comprehensive (i.e., it answers all engagement issues and expectations).
5. Release the final deliverable with enough copies for key client representatives, plus an electronic copy so the client can make more copies (but in a format such as PDF, which makes it more difficult to change the official report).
6. Schedule and hold final presentations and exit meetings.

to the board of directors, senior executives, and/or department and functional managers affected by the results of the engagement.

Deliverables also may include exit meetings with the management team, departments, and/or external stakeholders (such as community forums for a city or county client) to review the engagement's final product. These exit meetings are particularly important to ensure that the engagement steering committee, affected managers and staff, and stakeholders fully understand the substance of your deliverable and are committed to implementation or future actions. They also provide the basis for transitioning momentum and enthusiasm from you to champions and everyone else at the client. This activity may be your last chance to influence the client in person.

Key steps to ensure that the final deliverables are acceptable to the client are shown in Table 7.1. These steps are particularly important if the report will be released to the public or posted on the client's website (this is almost always the case in the public and nonprofit sectors, but also it is common in the private sector).

FINAL BILLING AND COLLECTION

Chapter 6, "Conducting the Engagement," provides advice on progress billing and collecting *during* the engagement. On the other hand, the final bill for the engagement should be presented with or just after the final deliverable. Follow-up at this step and throughout the course of the engagement is important to ensure quick payment. We recommend that the final bill follow-up should include the steps and responsibilities shown in Table 7.2.

Assuming you have presented a series of progress bills and the client has paid you promptly for these bills, particularly when the draft deliverable is billed, the final bill also should be processed in a timely manner by the client. A key step with the final, and every, bill is to check to ensure the client has received it, particularly if the bill was sent electronically.

Sometimes, the draft deliverable presents ideas or advice that are hard for the client to swallow, which can slow the engagement process down, including bill payment. Hopefully, you've softened the blow with periodic discussions of emerging

TABLE 7.2
Final Billing Follow-Up Steps

1. A follow-up e-mail or note to ensure the client's engagement liaison has received the bill and processed it for payment (after one week by the firm's project manager)
2. A second e-mail, or perhaps a letter, if the payment has not been received (after 30 days by the firm's project manager)
3. A third e-mail or letter, plus a phone call, if the payment has not been received (after 45 days by the firm's engagement director)
4. Implementation of the firm's collection process, if the payment has not been received or a firm payment commitment expressed by senior management at the client (after 60 days by the firm's engagement director and the firm's client relationship leader)

issues and preliminary ideas, which are early opportunities for you to sell the client on the need for something new and, perhaps, hard. For example, replacing or upgrading information systems, changing their approach to a long-standing service or product, and/or increasing costs rather than expected cost savings might cause the client to hesitate on going forward until they process and accept your ideas. If this happens, you need to be particularly attentive to the client at the presentation of the draft and final deliverables to ensure they are sold on moving forward. If not sold, the client can draw back, delaying progress on the engagement's final deliverable, as well as payment of bills.

FINAL REVIEW OF THE ENGAGEMENT WITH THE CLIENT

Once the final letter or report and presentation are delivered to the client, it is valuable to meet one final time with appropriate client personnel to go over the work performed. The firm's client relationship leader, who has ultimate responsibility for client satisfaction, should initiate this meeting within thirty to ninety days. Other attendees from the firm might include the engagement director and the project manager. Key topics to review are as follows:

- Did we achieve the client's goals for the engagement?
- Was our approach, work plan, and schedule appropriate and in conformance with the firm's proposal/contract (particularly if any items were changed during the engagement through contract amendments)?
- Were deliverables, including progress reports/meetings, draft deliverables, final deliverables, and final presentation, of acceptable quality and quantity?
- Was the client satisfied with the project as a whole and with our staff?
- Could we have done anything else to add value to the engagement process or deliverable?
- Is there more the firm could do to help the client?

This approach is important to gaining a solid reference and developing future work with the client.

Client:	Review Date:
Engagement Name:	Engagement Dates:
Engagement Budget:	Firm's Cost:
Client Representatives:	Firm Representatives:
Client Assessment 1. Process quality (approach and work plan) 2. Staff quality (director, manager, staff) 3. Schedule compliance (on time?) 4. Budget/cost compliance (on budget?) 5. Interim deliverable quality (progress/status) 6. Final deliverable quality 7. Benefit to the client 8. Overall client satisfaction	*Above, At, or Below Expectation (attach notes)*
Reference Name(s):	Phone Number and E-mail
Follow-On or Implementation Work:	Describe Below

FIGURE 7.1   Sample Client Satisfaction Interview Form.

You might want to prepare a simple client satisfaction interview form to document the closing interview discussed above. See Figure 7.1 for a sample form that can be filled in by hand (note that fields are compressed to save space in the book).

Gaining a Reference

Every engagement should result in a client reference that attests to the high quality of your work. At closure, you should ask the client to be a reference and determine who will be the specific reference listed in future proposals and qualifications statements.

All references should be confirmed annually, particularly to keep track of the contact person's latest job with the client and new phone number or e-mail address, or new contact information if he or she has a new employer. We regularly use people who have moved on in their career as references for past work. Sometimes these people gain additional industry status with the new job (e.g., a city manager leaving to head a national professional association).

Implementation Assistance

Many engagements offer recommendations or opportunities for future work that are beyond the agreed-upon scope in the original proposal/contract or because the engagement is Phase 1 of a multipart assignment, with Phase 2+ not yet fully defined.

Some implementation or Phase 2+ tasks may be work your firm can perform, while some may require the services of another professional service firm.

Care must be taken, however, not to make the initial engagement seem like a loss leader or a come-on for future work. The need for supplemental or implementation assistance work should flow naturally and appropriately from the deliverables for the current engagement.

However, it *is* appropriate to plant the seeds of follow-on or implementation assistance in the initial proposal and periodic engagement status meetings. For example, include a statement in the initial proposal that implementation assistance is not within the scope of this engagement but will be bid separately, if needed, by your firm or by appropriate technical experts at another professional service firm.

In our management consulting work, we often recommend facility improvements that would be developed by consulting architects and engineers, or identify the need for updated information technology that would be provided by outside hardware or software vendors. And at times implementation of some recommendations requires more in-depth or supplemental work by our firm to capture the full value from the recommendations in our deliverable, particularly if this work is outside the scope of the initial engagement.

SUBSEQUENT CLIENT FOLLOW-UP

As a part of the quality control process, it is good to call or meet with past clients nine to twelve months after the engagement is closed out, to ensure implementation or next steps are moving along as expected and/or the client is still satisfied with the deliverable. The update may indicate the need for additional assistance, if deliverables are resisted or proving hard to implement.

In our management consulting work, we often preschedule client follow-up either in the proposal or when closing the engagement. For example, employee surveys used on the original engagement may be reapplied twelve to eighteen months after engagement closure to measure positive movement in implementation, organizational culture, management philosophies, resource allocations, or personnel practices.

Accounting, auditing, financial planning, insurance, and tax preparation firms may work on a recurring (likely annual) schedule, with follow-up on past work and start-up for new work perhaps blending together into one client contact.

WRAPPING UP ENGAGEMENT DOCUMENTS AND FILES

Chapter 2, "Developing a Culture of Internal Discipline and Control," provides guidance on documentation standards; Chapter 4, "Setting Up the Engagement," describes the elements of engagement documentation (the engagement file and the working file); and Chapter 6, "Conducting the Engagement," covers how the engagement documentation should be managed as the project unfolds. This advice should ensure that your documentation is relatively complete and in order at the end of the engagement. But often, particularly in busy times at the end of the engagement, the work papers can become a mess—with stuff in boxes, in piles on the floor, on your desk and credenza, and maybe even at home (or in your basement if you have a home office). Right after the final deliverable is sent to the client is a great time to bring order from this chaos.

So now that you've finished the engagement, what are you going to do with this box (or all these boxes) of stuff that hopefully is now in order? One answer for a sole practitioner is to take it to the basement and find a corner to hide it. A better way is to conduct one last review of what you have and decide what's to be done with it based on the following steps.

1. Ensure that all key documents are in the file, including the request for proposal or client letter seeking assistance, proposal and contract, progress reports, bills, and draft and final reports or deliverables.

2. Ensure that all interview and discovery notes, documents and reports, budgets and financial reports, asset lists, statistics, and the like are included in the file. Catalog this information by organizational unit, topic/issue, or some other structure that makes sense related to the engagement's scope and issues. Ideally, information to support a specific deliverable should be easily accessible in the working papers.

3. Ensure that all analyses are included or identified in the file (some may be archived on a CD, a thumb drive, or in the cloud as computer files).

4. Determine what doesn't need to be retained in the documentation or should *not* be retained. For example, we keep draft deliverables for six months and then toss them out unless there is a strong reason to keep them longer (like someone publicly claims we just incorporated management's comments and changes in the final deliverable as our ideas). We'll retain indefinitely, however, electronic copies of the draft deliverables on our computer or on the *prior engagements* thumb drive.

   As a precaution, we also go through the work papers and cull out anything that might present a confidentiality risk to the firm or to client personnel, including interview notes, employee surveys, and employee questionnaires that have pointed or personal comments, or information that might hurt someone or cost them their job. The client culture during the engagement will tip you off to the need for aggressive culling.

   We typically keep the work papers or boxes for five to ten years and then shred them when that time is up if the basement gets overloaded with work stuff. The nature of your firm and your work will drive how long you keep work papers and what you keep.

5. Update your firm's qualifications statement, clients/engagements lists, and resumes; prepare a one-page engagement summary for the project; and file the final report in your deliverable file (to document all the work you've done over the years in one, easy-to-find place like the four-drawer file cabinet in the basement or back room).

INDEPENDENT EVALUATION OF THE ENGAGEMENT

Because we have emphasized the importance of a culture of internal discipline and control throughout the book, you should close your work with a client by reviewing adherence to your firm's IDC practices during the engagement. Ideally, someone should conduct this review who is independent of the engagement—not the engagement director or project manager. Some questions the independent reviewer should seek answers to are as follows:

- Was the client completely satisfied with the results of the engagement?
- Was a high quality of service and client care provided throughout the engagement?
- Is each of the deliverables specific enough to ensure complete client understanding of the nature and intent of the proposed action?
- Are the deliverables practical considering the client's special situation?
- Have the economic/fiscal aspects of the deliverables been properly balanced with the desire to produce a "perfect" result?
- Has the entire engagement been conducted in a manner most likely to advance the knowledge and development of client personnel and the long-term success of the client?
- As a result of having completed the engagement, is there concrete evidence of increased organizational effectiveness, reduced liability or risk, or improved financial condition? In other words, did the firm make a significant contribution to the client?
- Has every relevant, specialized skill within the firm and its external resources (like subcontracted technical specialists) been brought to bear on the engagement?
- Was the strength of the firm's convictions (values and guiding principles) maintained as to the client's balanced, long-term best interests? In other words, did the engagement team keep to the firm's IDC practices, or did they cut corners?
- Was the firm's overall financial return satisfactory? How profitable was the engagement?
- Were engagement risks adequately identified and avoided or managed effectively?
- Have there been any "learning points" that could improve the methodologies used, could be applied in future similar engagements, and/or could enhance the firm's services and service levels?

A big firm will have people available and assigned to this type of review. Barry, for example, conducted 100 independent reviews of large and long-term management consulting and advisory service engagements for KPMG as a retired partner on a contractual basis. Smaller firms, however, won't find this independent review so easy to conduct, as it's possible no one in the firm is independent of the engagement.

Client:	Review Date:
Engagement Name:	Engagement Dates:
Client's Budget:	Firm's Cost:
Engagement Leaders:	Independent Reviewer(s):
Process Assessment 1. Was our approach and work plan effective? 2. Did we use the right tools and use them well? 3. Were we on time? 4. Were we on budget?	*Above, At, or Below Expectation (attach notes)*
Project Team Assessment 5. Did we assemble the right team? 6. How did the engagement director do? 7. How did the project manager do?	*Above, At, or Below Expectation (attach notes)*
Deliverables Assessment 8. How did we do on interim deliverables? 9. How did we do on final deliverables?	*Above, At, or Below Expectation (attach notes)*
Client Assessment 10. Was the project of benefit to the client? 11. Was the client satisfied? 12. Did we gain a reference?	*Above, At, or Below Expectation (attach notes)*

FIGURE 7.2   Engagement Review Form.

In that case, we suggest you enlist past colleagues, a mentor, or friend with a good knowledge of your work to discuss the project with the engagement team, and then comment on the above questions. In effect, you would make a final presentation to that person as if they were your firm's independent reviewer.

Like the Client Satisfaction form (Figure 7.1), you should complete an Engagement Review form (Figure 7.2) to document the results of the independent evaluation of the engagement. Again, fields on the figure are compressed to save space in the book.

TIPS FOR SOLE PROPRIETORS AND SMALL FIRMS

Our formal process for closing the engagement might seem a bit much for small businesses or for sole proprietors. Nevertheless, it is important to incorporate these key steps into your business processes:

1. Get formal client approval for the final draft deliverable before you go final. There's little worse than having to pull back a "final" report to fix or update something! If you get the OK and they ask you to fix something, you can bill them for the repairs.
2. Have someone independent (not on the project) review the pre-release final deliverable. For example, engage—with compensation—a local author or a

retired friend to go over the final draft of a report before it is released ... or maybe your sister, if she is extremely picky.

3. Invest time in exit meetings with client managers and staff after the formal, final presentations to get their perspectives on your work and deliverables.

4. Be a pit bull in billing and collection of the final bill.

5. Wait a month or so, then make at least a phone call (or better yet hold a meeting) with the client to review their overall satisfaction with the deliverables and process, as well as to nail down the reference and explore opportunities for additional work.

6. Call back in six to twelve months to see how they are doing with implementation. Maybe they need more help.

7. Clean out the working papers, particularly to eliminate sensitive materials.

8. Update your qualifications statement, client/engagement list, and resume.

9. Find someone independent, trustworthy, and encouraging to review your work on the engagement and in general. Perhaps create an informal board of directors to give you advice and encouragement. Develop a mentor with professional service experience.

8 Applying the 9 Rules

This chapter of the book again lists the 9 Rules for Success and then provides two case studies on client engagements that successfully applied the rules and resulted in positive reviews for the engagement team and long-term references for our work.

9 RULES FOR SUCCESS

The 9 Rules are as follows:

Rule 1 Clearly define your market niche (industry or industries, geographical coverage, client size, and list of services) to create a unique and powerful offering to potential clients.

Rule 2 Develop and implement a system and culture of internal discipline and control to ensure consistency of service, efficiency of operation, and quality and reliability of deliverables. Then train, mentor, and monitor personnel regarding engagement management policies and procedures.

Rule 3 Establish and enforce engagement documentation standards, including those for proposals, progress reports, and deliverables.

Rule 4 Practice what you preach regarding internal culture, policies, procedures, and standards.

Rule 5 Prepare complete and definitive service proposals, contracts, and engagement work plans that evaluate and accommodate engagement risks for both the provider and the client, so the client knows what can be expected in terms of scope, work plan, schedule, deliverables, and cost.

Rule 6 Be flexible and adaptive to the "real world" once the project starts, to manage the dynamic between client expectations and what's *really* happening within both the firm and the client.

Rule 7 Implement a firm–client communication plan that will ensure clear and frequent discussion of engagement progress and status.

Rule 8 Bill and collect frequently. It will improve cash flow and alert you sooner if the engagement is in trouble with the client.

Rule 9 Conduct independent status/quality reviews of the engagement while it is in process and subsequent to conclusion. Involve key client contacts in the reviews.

CASE STUDY 8.1
Milwaukee Department of Public Works

From 1989 to 2001, Steve Egan conducted nine studies to improve the management, organization, operations, and resource management of the Milwaukee Department of Public Works and several of its operating bureaus/divisions (Administration, Engineering, Forestry, and Water Works).

CITY OF MILWAUKEE IN 1989–1990

The City of Milwaukee had a population of almost 630,000 in 1990 (a decline of 21 percent from its highest population in 1960). It was (and still is) governed by a mayor–council form of government. The city government was characterized by

- *Politics:* Like many northern big cities, the Milwaukee city government was very political, including politically oriented department heads. Elected officials and citizens expected a lot from the Department of Public Works, and the commissioner and his staff were under continual pressure to deliver on large projects and services (such as major road and building projects) as well as small ones (like repairing a raised sidewalk panel in front of someone's house).
- *Unions:* Almost all city employees were members of strong and vocal unions. Primary exceptions were senior managers, such as department commissioners and division/bureau directors. More than a dozen different unions represented Public Works employees. One five-person water crew had four different unions representing its workers.
- *Strong Central Support Services:* Support departments—such as budget, IT, and personnel—were professionally staffed and exerted strong influences on department operations and resource management practices (e.g., budgets, financial management, accounting and cost accounting, hiring and personnel actions, labor relations, and public information).

In 1989 the Department of Public Works (DPW) had over 2,650 employees and an annual operating budget exceeding $150 million, plus over $45 million in capital projects. Department facilities included thirty-nine yards, plants, and offices across the city.

The department was directed by a commissioner who was both technical (a licensed, professional engineer) and political (a mayoral appointee). The Office of the Commissioner had fewer than ten positions, several of which were administrative support positions.

DPW's nine bureaus (General Office, Bridges and Public Buildings, Engineers, Forestry, Municipal Equipment, Sanitation, Streets and Sewers, Traffic Engineering and Electrical Services, and Water Works) were highly independent.

Steve Egan's DPW Projects

Steve's work started and continued in the context of the new mayor's goal to lower the tax rate by increasing the tax base and cutting costs. These goals were complicated by "the insatiable appetites for additional funds of the Police and Fire Departments" (per the former budget official). A recurring theme was that DPW, because it had the largest budget, must cut costs to free up funds to meet the mayor's overall goals and to fund police and fire. This cost cutting process was hard, but in many ways rewarding to DPW in terms of efficiency, effectiveness, and stakeholder and customer service—along with a strong commitment to treating employees equitably. What follows is an overview of Steve's nine consulting projects.

Organizational Analysis

Steve's work in Milwaukee commenced in 1989 with an organizational analysis of DPW's senior management and bureau-level organizational structure. The project was one of Mayor John Norquist's initial efforts to restructure city services and cut costs after taking office in 1988. He followed the legendary Henry Maier, who had served as mayor for twenty-eight years.

This project reviewed the organization plan and management/supervisory staffing levels for the department and its bureaus. The goals were to redistribute activities, particularly for Engineering; simplify the department to improve accountability, communications, and reaction times; reduce levels of management; and reduce management and engineering costs. Recommendations and the implementation plan were intended to minimize the effect on employees and facilitate acceptance of recommendations.

The study, in concert with a companion analysis by the Budget Office, resulted in a savings of 141 management and supervisory positions without layoffs and annual savings of over $6.6 million. The study also restructured the management plan in each bureau and created new resource management staff in the Office of the Commissioner to improve the commissioner's ability to direct what had been *very* independent bureaus. One of the bureaus, Engineering, was headed by a mayor-nominated and city commission-confirmed city engineer, who was as politically powerful as the Commissioner of Public Works to whom he reported operationally.

Bureau-Level Functional Studies

The *Forestry Bureau* study in 1992 and 1993 was initiated in response to the mayor's plan to restrict resources to this operation, while requiring Forestry to continue to maintain the city's exemplary streetscape of trees, boulevards, and pocket parks. The study recommended increased cross-utilization of arborists and gardeners (a major cultural change that was highlighted in Mayor Norquist's book, *The Wealth of Cities: Revitalizing the Centers of American Life*),[1] continued operation of the city's nursery and greenhouse, and a range of operational improvements.

The *Engineering Bureau* study in 1993 was in response to the mayor's and city commission's question "What are all these 275+ engineers doing?" We coached the city engineer and his management team through an analysis of organizational structure, reporting relationships, and interdivisional cooperation and coordination. The study resulted in a reduction of ten engineering management positions; reassigned the assistant city engineer to an operational role leading the administrative section;

and enhanced interrelationships among related services (e.g., water and sewer engineering) and operating bureaus (e.g., Water Works).

The three *Water Works Bureau* projects started just after the spring 1993 cryptosporidium water quality crisis. Steve was tasked with identifying and resolving management, operations, and nontechnical process issues that impacted water quality and quantity, and impeded the Water Work's response to the crisis. Technical and engineering issues (such as relocation of the water intake "crib" in Lake Michigan away from the outfall of the Milwaukee River) were addressed by a team that included City Engineering, Water Works, City and State Health Departments, and the Federal EPA.

Concurrently, Barry Strock Consulting Associates (BSCA) was working to define utility management information system needs and acquire a vendor software package. Once these initial projects were completed, Steve teamed up with BSCA to conduct thirty-person, day-long team-building seminars for all 451 Water Works employees. These seminars identified and measured (on a 1–5 scale) employee *GRIPES*, which were issues related to *g*rowth and employee development, *r*espect and *r*ecognition, *i*nformation, skills and *p*otential tapped by the organization, *e*mpowerment, and *s*upport, both internally and externally.

A year later Steve served on the City/DPW Committee to recruit a new Water Works Superintendent.

Reengineering and Strategy Work

Steve's *Administrative Services project* involved preparing DPW for reengineering of all administrative functions in late 1996 and early 1997, as well as implementation of a citywide financial management information system in late 1997 and early 1998. In effect, Steve was imbedded into the DPW on a part-time basis for about six months, as he worked with staff across the department to reengineer various processes and prepare for system implementation. The engagement defined department-specific system requirements, reviewed and recommended vendor solutions, and developed a departmental implementation plan for the new financial management information system.

Steve worked with DPW in 2000 on a *policy analysis* to review job classifications, compensation issues, and business practices to facilitate the most efficient utilization of employees. He also assisted the DPW's management team in developing a *financial and operations strategy* to react to budgetary restrictions placed on the department due to citywide financial issues and limitations.

To supplement the review of human resources and business practices, Steve facilitated a day-long *budget strategy workshop* for the DPW's senior management team. The workshop focused on the application of the principles of high-performance, competitive government to DPW's budget challenges. Many of these principles were first presented in Osborne and Gaebler's book, *Reinventing Government*.[2] Key elements of the workshop included a review of The Mercer Group's *1997 Privatization Survey*, to identify activities and functions that might be contracted out and to reconsider the department's mission, vision, and values.

APPLICATION OF THE 9 RULES FOR SUCCESS

Rule 1   Clearly define your market niche (industry or industries, geographical coverage, client size, and list of services) to create a unique and powerful offering to potential clients.

We won the first contract in 1989 because we had a strong market niche with local governments in Wisconsin. The firm of record, David M. Griffith and Associates (DMG), had a twenty-year track record of service to state and local governments across the United States, which was strengthened by the hiring of a former state government official as the Wisconsin practice director—who was well known to City of Milwaukee decision makers. Steve's inclusion on the team as lead consultant further strengthened our position, due to the many similar studies he had performed in eight years of consulting in the public sector. So, our team was established, well known, mostly locally based, technically qualified, and experienced in the consulting work required.

Rule 2   Develop and implement a system and culture of internal discipline and control (IDC), to ensure consistency of service, efficiency of operation, and quality and reliability of deliverables. Then train, mentor, and monitor personnel regarding engagement management policies and procedures.

The IDC system for management consulting projects was an adaptation of the robust system that Steve learned and used when employed by KPMG and working for Barry Mundt. So, although DMG's IDC was oriented to its cost accounting work, the management consulting work still had a strong IDC foundation through Steve's training at KPMG. We applied this system on the first and later Milwaukee projects, which fortunately required small staffs of three to four people who were relatively easy to train, mentor, and monitor. Plus, we had enough time together on site to maintain cohesion in both the IDC and analytical activities. In other DMG assignments, however, teams with staff from multiple offices and different backgrounds/disciplines proved problematic; some team members were not trained in an IDC system for management consulting and/or had a different cultural background (accounting/auditing versus management consulting).

Rule 3   Establish and enforce engagement documentation standards, including those for proposals, progress reports, and deliverables.

Again, the standards connected back to Steve's days with KPMG and were tailored to the city's requirements, as expressed in the request for proposals (RFP) and the engagement kick-off meeting. For governmental work, the RFP typically describes in some detail the content of proposals, creation of a project steering committee, progress reporting expectations, project schedules, and the content of deliverables. Our proposal then worked these requirements into the context of our consulting methodology.

Rule 4 Practice what you preach regarding internal culture, policies, procedures, and standards.

DMG's Wisconsin practice director had a strong background in IDC from his work with the State of Wisconsin's Revenue Department. He had stature within DMG, personal integrity, and a commitment to applying a strong IDC system that led to client satisfaction with our work. As the fulcrum of the first three projects, he brought the cost accountants and management consultants together. Later projects were Steve's responsibility.

Rule 5 Prepare complete and definitive service proposals, contracts, and engagement work plans that evaluate and accommodate engagement risks for both the provider and the client, so the client knows what can be expected in terms of scope, work plan, schedule, deliverables, and cost.

Our initial organizational analysis responded to a formal public-sector procurement process, including request for proposal, pre-bid conference, proposal, on site presentation to review the proposal, contract, and engagement work plan and schedule. We accommodated engagement risks, which were as much political as process, by continual communication with the project steering committee, which included the city's budget director and DPW budget analyst, as well as the DPW's management team.

Like the 1989 DPW organization study, the forestry engagement followed a formal procurement process, with request for proposals, proposal, presentation, contract, and the like. To strengthen the project team, the budget office merged two proposers to take advantage of the strengths of each. Engagement management also was formal, with a risk mitigation strategy and plan much like the DPW organizational study.

The engineering study started from a less formal procurement, with proposals from a few select firms. This project was more of a coaching/facilitation exercise with the city engineer and key managers, rather than a full-scope organizational and operations analysis. Progress/status reports, billing, draft report reviews, and presentations followed our standard practices.

The Water Works project was crisis-initiated (the 1993 cryptosporidium outbreak in city water) with our involvement, including the team-building seminars, negotiated rather than set up through a formal procurement process. We were on site for the engineering study and quickly transitioned to the Water Works study. Proposal, consulting methods, progress/status reports, billing, draft report reviews, and presentations followed our standard practice.

The reengineering and strategy work again were informal but included a short proposal and contract. Client oversight for reengineering work was primarily by the administrative services director, as Steve acted like an imbedded analyst/consultant.

Rule 6   Be flexible and adaptive to the "real world" once the project starts, to manage the dynamic between client expectations and what's *really* happening within both the firm and the client.

The "real world" in Milwaukee is politics. Early in the DPW organizational analysis, we learned that the interests and needs of the mayor and city commission (like most local governments) must be factored into our consulting work, but without compromising professional standards. A key adaptation was to listen to politicians, both in fact finding and when we developed preliminary findings and recommendations. In effect, we learned enough during initial interviews to tailor, but not change in substance, our recommendations to the political reality in Milwaukee. Then we used draft report reviews to sell our ideas to the Budget Office, the DPW management team, the mayor and senior staff, and the city commission (through the Public Works Committee).

Throughout our work, the "real world" also included a squeeze on DPW budgets and staffing, as explained earlier. This was our Milwaukee context. Other client engagements at this time were more balanced in squeezing and optimizing budgets and staff, with some local governments able to increase budgets to meet key service or capital needs.

The Water Works project also was a sharp and quick adaptation to the "real world." We were working in an Engineering Bureau conference room when a Crisis Response Team from DPW, the city and state health departments, and the federal EPA kicked us out because of the cryptosporidium crisis. As the response to the crisis unfolded, the DPW commissioner felt that there might be nontechnical issues limiting the effectiveness and efficiency of DPW's response. As a result we were engaged to look at Water Works management, operations, processes, systems, and controls.

Rule 7   Implement a firm/client communication plan that will ensure clear and frequent discussion of engagement progress and status.

In every DPW project, we established an engagement steering committee to review our work and comment on draft deliverables. In addition, we informally communicated with key client personnel (the project liaison, budget director, DPW commissioner, and bureau directors) on a regular basis. Because we had small project teams

(three to four) and a small group of client personnel to work with, our communications plan was relatively simple, but complied with our internal proposal and engagement management policies:
- Establish an engagement steering committee
- Informally talk with key client personnel weekly
- Formally report progress and emerging issues at least monthly
- Bill frequently to identify any red flags or potential problems
- Use the draft report to build commitment to our ideas and recommendations among city staff affected by our work
- Brief the mayor and key city commission members on issues that are politically sensitive

Rule 8   Bill and collect frequently. It will improve cash flow and alert you sooner if the engagement is in trouble with the client.

Steve billed 10 percent at the engagement kick-off meeting, with several goals: get set up as a vendor in the accounts payable system; find out how long it takes from the bill to a check in the mailbox; and cover first trip expenses. Later, progress bills were issued at least monthly to cover hours and expenses incurred during the billing period. All bills included a short description of the work performed, work to be conducted in the next billing cycle, and budget/schedule status, to solicit client comments on our work to date.

In Milwaukee, more formal and extensive progress/status reports were developed for client progress/status meetings. Otherwise, the billing and collection process followed normal procedures.

Rule 9   Conduct independent status/quality reviews of the engagement while it is in process and subsequent to conclusion. Involve key client contacts in the reviews.

The quality reviews in Milwaukee were limited to the client and two key members of the consulting team: the project director and lead consultant. Key client contacts were involved in periodic (not less frequent than monthly) status/progress meetings, reviews of preliminary ideas, and reviews of draft and updated draft reports. These meetings and reviews are a point of emphasis for Steve, as they serve to warm up and soften up the client to our ideas and eventual recommendations. Basically, we try to sell our ideas step-by-step, as the engagement unfolds.

Each of the nine Milwaukee Public Works engagements was well received by the Budget Office, Department of Public Works, and the mayor and city commission (when involved). The sign that we had successfully applied the 9 Rules was the city's continual calls for our services over a dozen years, development of several strong references, and partnerships with city staff when they retired or moved into management consulting.

CASE STUDY 8.2
International Firm

Frank Smith conducted multiple engagements for an international firm. The engagements concerned the evaluation of business processes and the application of technology in various operations. Several of these projects are explained in this case study, along with Frank's application of the 9 Rules.

ORDER AND WAREHOUSE MANAGEMENT SYSTEMS INTEGRATION/IMPLEMENTATION PROJECT

An independent consulting firm engaged Frank to conduct a training session and perform an assessment to determine the potential for deploying bar code technology in the warehouse at its client. The engagement allowed for three weeks of work. During the assessment, Frank discovered that the client had already implemented a bar code system in its warehouse. He also learned that the client was getting ready to implement a new warehouse management system, together with an order management system. Frank managed to get himself invited to the implementation team meeting to make observations.

Frank discovered that the client was preparing to go live with the implementation and integration of an order management system and a warehouse management package. The project was six months behind schedule and approximately $800,000 over budget. Each application had been developed and sold by two different software vendors. The order management system vendor did not have warehouse management capability as part of its product offering and so teamed up with another vendor. Further, the software vendor had engaged a third-party consulting firm to manage the implementation.

Frank asked the director of operations about the process that his company used to select the software combination. The director said that the information technology department had made the selection. The operations group had no role in the selection process. The order management system salesman told the client that it had a "synergistic relationship" with the warehouse management system vendor. Frank did not understand what those words meant, so he inquired of the order management vendor where this software duo had been implemented previously. He discovered that the client was the first company to implement this software combination. Further, he learned that the integrated package had not been tested, nor had a software test plan been developed. A cursory inspection revealed that the company's operations would be put at high risk if they proceeded with the implementation.

Frank went to the newly hired vice president of operations and presented his findings. When asked what he thought should be done about it, Frank recommended that the implementation be stopped immediately. He reasoned that the team needed to step back, regroup, and chart a course that included the development of a risk mitigation strategy and plan. The vice president concurred and announced an immediate halt to the project one week before it was scheduled to go live.

Scope of Work

The scope of the project was to reconstruct an implementation team, construct an implementation strategy and plan, and execute the plan to completion.

Approach

The vice president of operations assigned Frank to lead the implementation team. Frank formed a new implementation team; established a go-forward strategy; and assembled a project plan with specific activities, deliverables, responsibility assignments, and due dates. He also constructed a risk mitigation strategy and plan to identify what could go wrong, detail the steps to take to prevent the identified event from occurring, and define steps to take if an event occurred despite best efforts to prevent it from happening. Finally, a communications strategy and plan was assembled to keep the company at large informed as to what was happening on the project.

The implementation team proceeded with the following phases:

- Facilitated working sessions to document the *as-is* and *to-be* process flows that the system was expected to support
- Constructed test case scenarios to validate that the system would be able to perform each action and produce the expected outcome
- Assembled a conference room pilot (set up a computer lab with a separate test instance of the software and database to execute the software test plans)
- Executed the software test plan (performed iterative positive, negative, volume, regression, and full system test scenarios for each of the test plan steps)
- Launched corrective action programming for each anomaly or unanticipated result that occurred and retest
- Confirmed the execution of complete full system test, without incident
- Assembled *go-live* plan that included a fallback scenario (part of the risk mitigation strategy and plan)
- Executed the *go-live* plan

Results of the Engagement

Successful *go-live* was achieved twelve weeks from the date the original project was stopped. Numerous system anomalies were identified during the conference room pilot. Going live with the original warehouse management system would have been catastrophic had the company not executed the software test plan and corrected the problems that were identified.

FIVE-YEAR OPERATIONS PLAN REVIEW

During the order management and warehouse management system implementation, Frank learned that the client had prepared a five-year operations plan to support its goal of doubling its size in five years. Frank asked the vice president of operations if he had conducted an assessment to determine the plan's impact on the operations department and its ability to support projected growth (of 100 percent) over five years. An assessment had not been done, and the vice president engaged Frank to conduct the assessment.

During the assessment, Frank discovered that the revenue forecast was not supported by detail volume data from any of the program departments; rather it was an extrapolation based on current volume. Findings and recommendations were submitted in a written report to the vice president of operations. The recommendations included four technical enabling initiatives that were deemed to be critical to the operations department and the company.

Identification of plan shortcomings (lack of detail support data) highlighted the need to develop a forecasting capability. When the findings describing how the revenue forecast was determined reached the office of the president, he called the senior staff together and directed the heads of the various divisions to redo their forecasts and base them on hard data. Frank also recommended development of specific management skills in order to realize operations department improvements in efficiencies and utilization.

Scope of Work

The scope of the engagement was to do the following:

- Evaluate the impact of growth on the operations department
- Determine the actions that may need to be considered to accommodate anticipated growth
- Determine growth impact on management needs and skills for operational areas

Approach

Frank met with program and department leaders to review revenue growth reflected in the plan to

- Identify supporting schedules for the plan
- Identify products and services contained in supporting schedules
- Determine products and services that would impact operations
- Attempt to determine the impact of anticipated revenue growth on operations functions
- Identify and highlight technology initiatives that had the potential of impacting operations in a positive way

Then, first-round interviews were conducted with program leaders to discuss the five-year plan; confirm that the revenue projections were accurate and had not changed; and determine progress to date, assuming a linear growth curve.

Frank met with directors, managers, and supervisors, in each of the functional areas of the operations department:

- Call Center
- Distribution and Logistics
- Planning and Account Management
- Publishing and Packaging
- Quality and Operations
- Other Services

Results of the Engagement

Frank identified specific actions that needed to be taken by the operations department to support corporate growth goals and refocused efforts to implement all of the necessary actions. The company achieved its growth goal.

In addition, Frank was retained to lead the implementation of several technology initiatives, all of which were implemented successfully. These initiatives required the independent firm that Frank contracted with to bring in additional specialized resources that included the following:

- A team to design and implement a fully automated material handling infrastructure
- A team to train and implement Six Sigma Lean management practices
- A team to manage the selection of custom-designed automated material handling equipment

SELECTION AND IMPLEMENTATION OF AN IMAGE MANAGEMENT SYSTEM

Market forces, together with the company goal to double its size in five years, established a need to develop an alternative process for storing large volumes of printed materials. The company was in the process of selecting an Image Management System (IMS). However, responses to the requests for proposals (RFP) were unsatisfactory, and it was unclear as to whether options being considered would satisfy company needs. Frank was asked to intercept the IMS selection process and get the project back on track.

Scope of Work

Frank facilitated several working sessions with key stakeholders to define the business processes that the software needed to support. The completion of the exercise allowed the team to refine its project and requirements definition. A review of the RFP that was sent out revealed that it did not address all of the requirements that were defined as output to the exercise above. This led to the decision to send out an addendum to the original RFP.

The original responses, together with answers submitted by vendors to the addendum, were subjected to a structured evaluation. The evaluation led the client team to make a final selection with a high level of confidence that it would satisfy their needs, and where requirements couldn't be satisfied, acceptable workarounds would be established.

Finally, the team entered into contract negotiations that included a variable fee structure based on use, which mirrored the peaks and valleys of the company's workload profile over the course of a year.

Approach

Frank assumed the role of overall project manager of the IMS implementation and introduced project management discipline, tools, and techniques. The team developed a way to quantify forecast volume by digital attribute, as well as related scanning volume to determine required scanning capacity.

He assembled a project team, documented the system flow design, and used the flow design to evaluate the adequacy of RFP questions. The team developed a supplemental RFP and distributed it to vendors. Frank introduced a structured evaluation process to enable the team to make a final selection. A risk mitigation strategy and plan was prepared and followed to ensure on-time completion of the project. The team also prepared documentation to reflect the vendor selection rationale.

Results of the Engagement

The team selected an Image Management System and achieved *go-live* in time for the peak volume periods. The project was completed at more than $100,000 *under* budget.

APPLICATION OF THE 9 RULES

Rule 1 Clearly define your market niche (industry or industries, geographical coverage, client size, and list of services) to create a unique and powerful offering to potential clients.

Frank Smith's market niche included extensive experience with inventory management and control concepts, bar code systems training and implementation, and technical enabling initiatives that included requirements definition, evaluation, and selection. This market niche allowed him to be considered and to gain entry to a significant client.

The market niche was defined not by the industry served but by the methodology used for selecting and implementing technical enabling initiatives that were needed to solve business problems or enhance a business process. Frank was not well known by either the client or the primary contractor. His niche was providing services to a vendor who did not have bar code knowledge and skills within its ranks and/or lacked available resources to deliver the services.

Rule 2 Develop and implement a system and culture of internal discipline and control, to ensure consistency of service, efficiency of operation, and quality and reliability of deliverables. Then train, mentor, and monitor personnel regarding engagement management policies and procedures.

The IDC system for management consulting projects was an adaptation of the Runaway Systems project management methodology and Systems Development Life Cycle concepts developed first at Peat Marwick, Mitchell & Co-partners (now KPMG). Frank learned and used these tools when employed by KPMG. Frank was the only consultant from the vendor firm working on this part of the engagement, so he trained client personnel on these methodologies.

Rule 3 Establish and enforce engagement documentation standards, including those for proposals, progress reports, and deliverables.

A proposal had already been executed and other aspects of the engagement were well under way when Frank arrived. Nevertheless, Frank adhered to a cardinal rule of professional service delivery: manage client expectations. This was accomplished by focusing on the aspects of the engagement that related to his scope of work. He put a finer point on the definition of the deliverables by being crisp about what needed to be done and how it was going to be accomplished. Frank provided frequent updates on progress and findings, and developed short, but focused, presentation documents to management.

While completing his first three-week assignment, Frank learned that the company was preparing to go live with a new order and warehouse management integrated system. He heard some rumblings from the consulting firm managing the project that it was not going well. Frank asked to be invited to the next progress meeting just to observe. The consulting firm managing the implementation was lamenting the fact that they were preparing to go live, even though the system hadn't been tested. Aside from lacking a software test plan, no documentation existed as to exactly what the software was supposed to enable the company to do. That meant that there was no standard on which to measure whether the system was going to enable them to do what they were expecting it to do and no plan to test it. This situation presented an opportunity to provide additional professional services.

Rule 4 Practice what you preach regarding internal culture, policies, procedures, and standards.

The history of technology engagements is that they often go over budget and schedule, and fail to meet expectations. Consequently, they have a reputation of being high-risk ventures. Frank's experience is that companies fail to follow a structured process when selecting technical enabling initiatives, whether they are targeted solutions or enterprisewide applications. Following a structured process allows the stakeholders to identify gaps that the selected solution will not satisfy. This allows the team to be proactive in developing workarounds to fill the gaps.

Rule 5 Prepare complete and definitive service proposals, contracts, and engagement work plans that evaluate and accommodate engagement risks for both the provider and the client, so the client knows what can be expected in terms of scope, work plan, schedule, deliverables, and cost.

Each of the client's projects highlighted above lacked work plans and did not accommodate risks. The $800,000 cost over budget was evidence that the project was not being managed well. Creation of a risk mitigation strategy and plan would have increased the probability that problems and solutions would have been identified earlier in the engagement. A risk mitigation strategy and plan were assembled for each of the initiatives.

Rule 6 Be flexible and adaptive to the "real world" once the project starts, to manage the dynamic between client expectations and what's *really* happening within both the firm and the client.

Many companies employ very accomplished employees who are knowledgeable about the services they deliver. What is often lacking is documentation about the way they are expected to go about delivering them. Oftentimes, management assumes that the workforce has the skills that are needed to perform a function. In the case of the five-year operations plan, Frank started by identifying the skills that the operations department needed in residence to support the business growth that management had charted. He then created a Skills Flexibility Matrix (see Figure 2.3) to determine the gaps that existed between the skills that were needed versus those that resided in the organization. The Skills Flexibility Matrix was the basis for creating a training program that focused on eliminating the skills gaps that were identified.

Although a risk mitigation strategy and plan had been developed, nothing had prepared the team to accommodate a major event that occurred during the engagement. Emotions waxed and waned over the weeks and months that followed. Frank is convinced that the structure and plans that were put in place guided the team to complete its mission, even on days when everyone seemed to be in a fog.

Flexibility became the order of the day. The major event required the client to alter the primary way that it had been conducting business for years. The image management system became a primary enabler that supported a revised critical business process.

Rule 7 Implement a provider/client communication plan that will ensure clear and frequent discussion of engagement progress and status.

Preparing a communications strategy and plan is directly related to the cardinal rule: manage client expectations. The team defined messages that needed to be communicated, the intended audience, the frequency, and the mode, and then confirmed that the messages which were sent were the messages that were received.

The team also embraced an implicit component in the communication strategy: professionals do not need to wait for a regularly scheduled meeting to communicate an unexpected event. Managers dislike surprises. Communicating discrepant events as soon as they became known reinforced the client's perception that the team could be trusted.

Rule 8 Bill and collect frequently. It will improve cash flow and alert you sooner if the engagement is in trouble with the client.

Billing and collection begins with the engagement contract. The contract should be clear and unambiguous in stating the fees, if expenses will be included, how often the client will be invoiced, the terms, the amounts, and what will be included in each billing. Even though the primary contractor handled billing and collection, Frank played a role by "keeping a finger on the pulse" of aging receivables. He always makes it a point to meet the accounts payable person handling payments and cultivating a positive relationship with him or her. If an expected payment is late, a quick visit to that person will usually reveal who is causing the delay in the invoice approval process. It also helps to learn the schedule for when checks will be run and exactly how the client's internal approval process works.

Rule 9   Conduct independent status/quality reviews of the engagement while it is in process and subsequent to conclusion. Involve key client contacts in the reviews.

The primary contractor conducted face-to-face status meetings with the client on a regular basis. Frank was intimately involved in this process to ensure that neither the client nor the primary contractor was on the receiving end of surprises.

WRAP-UP

We hope our explanation of the steps in developing and applying a culture and system of IDC in professional service delivery—as well as our two case studies, proposal preparation guide, and report (deliverable) preparation guide—will be beneficial to you and your colleagues as you start and/or grow your professional service firm. We emphasize once again that the keys to your success include a strong and consistent commitment to IDC.

ENDNOTES

1. Norquist, John O. 1998. *The Wealth of Cities: Revitalizing the Centers of American Life.* Page 32. Boston: Addison-Wesley Publishing Company, Inc.
2. Osborne, David, and Gaebler, Ted. 1992. *Reinventing Government: How the Entrepreneurial Spirit Is Transforming the Public Sector from Schoolhouse to Statehouse, City Hall to the Pentagon.* Various Pages. New York: Addison-Wesley Publishing Company, Inc.

Appendix A: Proposal Preparation

Proposals, along with requests for proposals and sample contracts in the public sector, are the foundation for defining professional services to be provided to a client. This appendix provides a guide to assessing a business opportunity and to preparing and presenting proposals to a prospective client, including these four topics

- Pre-Proposal Assessment
- Proposal Preparation
- Proposal Presentation
- Post-Proposal Follow-Up and Review

Each of these topics is described in the following pages.

PRE-PROPOSAL ASSESSMENT

Before you start to write a proposal, you must understand enough about the client and the client's need for a professional service, such that you are able to state explicitly what you intend to do for the client, how you are going to do it, what you expect the client to do to support your engagement, and what the engagement will cost (to ensure a competitive price for the client and a reasonable profit for the firm). This understanding can be developed through the following:

- Research on the client and its industry
- Interviews with the prospective client's senior executives
- Discussions with senior members of your firm who may have insights on the client

The assessment should define any potential problems or risks with serving this client, such as follows:

- Conflicts of interest (e.g., personal ties between people in your firm and the prospective client that might affect independence and objectivity)
- Business risks in serving the prospective client (e.g., if the client is suspected of doing something illegal or an engagement may open your firm to a potential lawsuit)
- Unacceptable client expectations, such as a significant work effort for a small price or at reduced rates (they're cheap) and/or a too-compressed project schedule that poses risks to service or deliverable quality (they're in too much of a hurry)

- Lack of key skills and experience in the industry or technical project issues that might put your ability to meet client needs and your reputation at risk (e.g., lack of information technology experience in the specific application of interest to the client; experience in commercial but not residential work)
- Current workload and resource availability does not allow you to successfully serve the prospective client

These and other risk factors should be enumerated, documented, and taken into consideration as part of the pre-proposal assessment.

Because the proposal often becomes a legal contract once written acceptance is acknowledged by the client or is included as a key element of a formal contract document, it is critical that you carefully define the engagement's objectives, scope, work plan, schedule, and price as part of your risk mitigation process. If you have doubts about the client or your ability to successfully complete the engagement, don't propose! The longer you work in the professional service field, the better your gut will tell you if you should propose or not propose.

PROPOSAL PREPARATION

In effect, the proposal should represent a formal meeting of the minds between the professional service firm and the client. To accomplish this level of agreement, an initial proposal may be reviewed and amended before a final proposal is accepted or the engagement starts and, rarely, during the engagement if additional needs are identified by the firm *and* the client.

A proposal can be a short and simple letter or a complex document. In rare cases, the proposal is a confirming letter that documents what has been agreed to verbally, such as for add-on tasks for an open engagement. A well-defined proposal will include the following elements, with proposal amendments covering any of the elements needing revision. The order of these elements may change from proposal to proposal, particularly if a potential client issues a request for proposals (RFP) that defines the content and sequence of information required in the proposal.

- Title page
- Cover letter
- Introduction
- Background to the engagement
- Objectives of the engagement
- Scope of the project
- Deliverables
- Approach
- Work plan
- Schedule
- Progress meetings and reports
- Firm qualifications
- Project team qualifications
- Client participation
- Benefits
- Costs
- Conclusion

Each of these proposal elements is discussed below. The examples given are drawn from a variety of client situations. Examples of the structure of a legal proposal for bond counsel work and a management consulting proposal for a Parks and Recreation study are included as Tables A.1 and A.2, respectively.

TITLE PAGE

If the proposal is bound, the first page typically is used as a title page, which includes the name of the client, project/engagement name, the words "Draft Proposal" or "Proposal," the name of the firm, address of the firm, phone number(s), e-mail address, website, and date.

COVER OR PROPOSAL LETTER

If a letter proposal is submitted, the letter will contain the entire proposal, including all of the above elements, but in abbreviated form.

In a more comprehensive and lengthy proposal with multiple chapters, sections, appendices, and attachments, the cover letter will open the proposal, provide an overview of key proposal strengths, and acknowledge compliance with RFP requirements (if needed). A detailed table of contents may follow the cover letter, or a list of key proposal sections, with page numbers, may be imbedded in the letter.

A separate transmittal letter may be prepared if the firm is sending multiple copies of the proposal and, perhaps, supplemental information. The transmittal letter identifies the specific documents provided in the proposal packet (e.g., five copies of our technical proposal, one copy of our cost proposal, and an attachment containing our complete qualifications statement.

INTRODUCTION TO THE PROPOSAL

The introduction can be a stand-alone section of the proposal or one part of an Overview of the Project section that contains the introduction, background, objectives, scope, and deliverables. Regardless of the proposal format, the introduction should contain statements relating to the circumstances giving rise to the proposal opportunity and what the proposal intends to do for the client. For example:

> (Name of firm's contact person) recently met with (name of client contact) to discuss how our firm might be of assistance in advising your senior executives on financial and retirement planning, and the company on succession planning. This proposal summarizes our understanding of the assistance you seek, outlines the nature and scope of our work, and sets out costs to conduct the engagement.

BACKGROUND TO THE ENGAGEMENT

The Background section should provide a brief summary of the events directly leading up to the situation at hand, contain pertinent additional information having a direct bearing on the situation, and provide the reasons why outside assistance is sought or being considered. For example:

Your management team has several key officials in sight of retirement. The company needs assistance in succession planning to develop or attract qualified replacements when the time comes for their retirements, and individual executives need assistance in end-of-career financial and retirement planning. Your Human Resources Department is not experienced in providing this advice internally, so you are seeking outside assistance.

OBJECTIVES OF THE ENGAGEMENT

A proposal may have one or several objectives that provide clear, measurable, and time-specific statements of what is to be accomplished by the professional service firm during the engagement.

At the end of the engagement, the client should be able to clearly see that these objectives have been met. For example:

> You expect our firm to meet with the following four key executives (list titles) and develop a personal financial and retirement plan for each within two months of project kickoff. In addition, you expect us to develop a succession plan for each of these positions based on the (name of firm) model discussed later in the proposal. This plan is to be completed within three months of engagement kickoff.

SCOPE OF WORK

The Scope of Work statement is one of the most important parts of the proposal, as it defines boundaries for the work to be performed and products or services to be delivered, thereby driving the project's cost. This statement should define the areas, activities, services, and/or functions to be covered and *not* covered by the engagement. For example:

> This project will be confined to the topics listed in the Background and Objectives sections of this proposal (financial/retirement planning and succession planning) for the four listed executive positions. The project will *not* review the compensation plan or the job performance of individual executives, nor will it involve planning for other executive positions.

ENGAGEMENT DELIVERABLES

The Deliverables section should clearly state what the professional service provider will deliver at the end of the engagement, which may include a written report on findings and recommendations, a summary of the results of meetings, technical drawings or exhibits, financial statements, deeds or other legal documents, representation in court, and the like. For example:

> At the end of the engagement, we will provide each executive a report that summarizes financial and retirement issues that need to be addressed, along with a road map on how to address them.
>
> We also will provide (name of company) a written succession plan for each position, to include key attributes required of the executive in the position, experience and training required to successfully fulfill job requirements, opportunities and strategies to develop internal candidates, and the need to look outside the organization to fill these positions.

Approach to the Engagement

The Approach section should clearly define how the engagement will be conducted, as well as specify the application of methodologies, tools, and techniques that will suit the client's needs and support the firm's standard methodology. This section sets up the detailed, step-by-step work plan that is contained in the following section. For example:

> We will apply our firm's time- and client-tested methodology for property inspections. This methodology includes assignment of a very experienced inspector, a 50-point checklist of areas to inspect, and a computer-assisted inspection report that is delivered to you immediately at the conclusion of the inspection visit.

Engagement Work Plan

The work plan is a step-by-step list of actions that are required to complete the engagement. Typically, the work plan starts with a kick-off or orientation task, during which the engagement's goals, work plan, schedule, and deliverables are confirmed or adjusted. Research and document reviews, however, might precede a kick-off meeting. Key work plan steps for many professional services include the following:

- Kick-off/orientation with the engagement's key client personnel or a steering committee to go over and confirm objectives, scope, approach and methodology, and work plan and schedule
- Research, document reviews, and industry comparative analyses
- Fact finding through interviews, site visits, and data/document collection
- Analysis of what's been collected during fact finding
- Development of emerging issues, ideas, conclusions, alternatives, and recommendations
- Preparation and presentation of engagement deliverables, typically starting with a draft report
- Review of the draft report and deliverables with key client personnel
- Preparation of an updated draft/deliverables and a second review with the client (in some cases)
- Preparation of a final report/deliverables, along with an implementation plan, if appropriate
- Presentation of the results of the final report/deliverables to the client, perhaps including the governing board and internal and external stakeholders

Engagement Schedule

The engagement schedule should define not only the start and end date for the engagement, but key milestones along the way, so the client and the firm can track progress and compliance with schedule and budget. Such milestones might include the following:

- Kick-off, if the date is estimable when writing the proposal
- Fact finding task start and end dates

- Progress meetings during the fact-finding task
- Analysis and draft report/deliverables preparation start and end dates
- Final report/deliverables delivery
- Final presentation delivery

It is important to ensure that the prospective clients know when the engagement can be started and how long it will take to finish. You should estimate a kick-off date in the proposal both (1) to assure the client that you are available relatively quickly after the proposal is accepted and (2) to manage your schedule for this and other client work. You might want to write this up as a tentative start date or an assumption that the client will approve the proposal within a certain time period (e.g., two weeks or a month).

Estimating the delivery dates of draft and final reports/deliverables is trickier, as several variables must be considering in developing the engagement schedule. Among these variables are holidays, the firm's commitments to other clients, the client's ability to quickly review draft deliverables, and the client's commitment to scheduling fact-finding meetings and site visits.

For example, the schedule blew up on one of Steve Egan's management consulting projects because the client's key draft report reviewer couldn't find the time to go over the results of a utility rate study. It took six months for him to get through the draft, which required a change order to update the data to the client's next fiscal year.

Progress Meetings and Reports

Because of the importance of *no surprises* during a professional service engagement, progress meetings and reports are critical to the success of the engagement and must be carefully defined in the proposal. Key means of communications include the following:

- Kick-off meeting
- Frequent information conversations with the engagement steering committee members and the clients liaison as the engagement unfolds
- Periodic (monthly at least) progress/status meetings
- Draft report/deliverable meeting(s)
- Final presentation(s)

The proposal should state the nature, frequency, and content of the progress/status reports and meetings. A typical progress report would include a discussion of what has been accomplished to date, what work is planned in the near future (up to the next progress report), what issues and challenges have arisen that affect the work plan and schedule, and emerging issues related to the technical focus of the project. For example:

We will prepare biweekly, written progress/status reports for your project director. These progress reports will include (list contents). In addition, we will meet with the Engagement Steering Committee monthly to review the project's status and emerging issues.

Benefits Received from Hiring Our Firm

All clients want to know why they should engage a professional service firm, so it's important to be as specific as possible in the proposal. But beware of overpromising! Typical client benefits are some combination of these:

- Delivery of a specific product (e.g., report, survey, inspection, facility design with plans) or service (e.g., representation in court, executive coaching, information technology needs assessment)
- Cost savings or revenue enhancement (in a consulting/advisory context)
- Process or efficiency improvement (in a consulting/advisory context)
- Better quality of products and services
- Improved competitive or market position

Two examples are set out below:

The Succession Plan will position the company to ensure that key management positions are filled with highly trained and experienced individuals who fit the company's organizational culture and management philosophy.

Our law firm will provide representation in court in areas where your internal legal department lacks functional knowledge and litigation experience. Specific benefits to (name of client) include representation by lawyers with over (number) years of experience in (the areas of interest), as well as representation of over (number) clients in litigation in these areas.

Firm Qualifications

The Qualifications section of the proposal should convince the potential client that you have the *experience*, particularly with similar clients on similar engagements, and the *skills* (i.e., methodologies, tools, and techniques) to successfully complete the engagement.

The Firm Qualifications section typically will include the following information:

- Introduction to the firm (date founded, history, number of offices and employees, services, industry or functional focus, and key national experience)
- Introduction to the local office (similar information to above, but with local office information and experience)
- Project experience that is similar or identical to the potential client's needs (often a list of ten to twenty engagements, perhaps categorized by type of service)
- References (four to six for similar clients, local clients, clients known to the potential client, and/or notable clients)

For example, the Firm Qualifications section of the proposal could include language similar to that adapted from a law firm's proposal for bond counsel services.

(Name of Firm) is a national law firm that specializes in state and local government finance. Our firm has been listed among the most active firms in the nation in every year since such rankings were first compiled in the mid-1980s for this type of work. We were ranked the number xxx bond counsel firm in the nation in 2012, based on dollar volume—higher than any bond counsel with a local office. Additionally, our firm was ranked the number xxx disclosure counsel in the nation in 2012 based on dollar volume—higher than any disclosure counsel with a local office.

As our proposal explains, we have experience in every area of public finance in which the (client name) has indicated a need. We have assembled a team of lawyers, led by lawyers in our local office, who can effectively and efficiently provide all the services the board would require.

In this state, our firm has been involved as counsel in many significant state and local public finance transactions and financing programs. Attorneys in the local office represent a wide range of governmental and nongovernmental clients in a variety of legal areas. Among the firm's local governmental clients are (list ten recent clients). Transactions for these and other local clients have involved virtually every type of municipal project conceivable.

ENGAGEMENT TEAM QUALIFICATIONS

A professional service is not only professional; it is personal. Specific members of your firm's team will be working with specific members of the client's team. Therefore fit and comfort among members of the two teams are critical in developing a relationship that leads to engagement success. Like the Firm Qualifications, the Engagement Team Qualifications section should convince the potential client that the team assigned to them has the experience and skills required to successfully complete the engagement.

After deciding to propose, the selection of the engagement team is your most important decision and challenge. As detailed in Chapter 4, "Setting Up the Engagement," your team could involve only one or two people, or a larger team that includes the following:

- *Relationship leader*: The person with overall and continuing responsibility for services to this client and ultimate client satisfaction.
- *Engagement director*: The person with overall responsibility for the quality of this specific professional service. Accordingly, the engagement director should have the technical and experience base to ensure the project meets both client and professional expectations. In a small firm, this person must be able to pull off the project all by himself or herself if everyone else fails or key people leave the firm.
- *Engagement manager*: The person with responsibility for day-to-day and step-by-step management of the project, work plan, schedule, budget, and deliverables. This person must be technically proficient *and* be able to manage the engagement team of 10 or more members.
- *Team leaders*: On larger projects, this person has responsibility for a segment of the work, such as a subdeliverable, functional area, or client site or department. Like the project manager, this person must be technically

proficient in that area *and* be able to manage his/her team of up to three to four professionals.

- *Team members*: Team members do much of the fact finding and analysis on a professional service engagement. They must be capable of effectively interacting with client personnel on interviews and of developing data and documents related to the engagements.

Steve Egan's approach to management consulting work, for example, is to bid on smaller projects that require only as many as four to five people, which may include an engagement director, the project manager, and two to three technical/functional consultants. On the other hand, Frank Smith's and Barry Mundt's work tends to involve larger teams, particularly for clientwide information technology projects that require a relationship leader, engagement director, project manager, and multiple team leaders with two- to three-member teams.

In any case, the makeup of the team and the presentation of the skills and experience of the team members are critical to a successful proposal. Although this section of the proposal will provide an organizational chart and short bios for *each* team member, an attachment may provide more complete resumes for at least the key team members, along with a list of their related past engagements. For example:

> The firm's senior professionals and affiliated technical specialists regularly team on strategic planning, management, operations, and resource management consulting projects. We believe that the combination of our firm's experience and skills in (list them for your firm), as well as our technical specialist's work experience in (list them for an associated firms or individuals), creates a team that covers all the bases for an engagement of this kind. Short bios for key team members follow.

Steve's short bio in a proposal for a Parks and Recreation Department Management Study, for example, reads as follows:

> Stephen D. Egan, Jr., a Mercer Group senior vice-president, will serve as our project manager and lead consultant. As project manager he will work with our project director to manage the day-to-day work of the project team, manage the project schedule and budget, and prepare and present progress reports and deliverables. As lead consultant, he will administer surveys, conduct most interviews, observe operations, and collect management, operational, and financial data to support the study.
>
> Steve is a former Fulton County, Georgia, Budget official whose responsibilities included analysis of department budget requests, internal consulting, and special projects for the County Manager and Board of Commissioners. He also served as interim Public Services and Water Director for the City of Highland Park, Michigan, under the direction of a state-appointed Emergency Financial Manager. Responsibilities in Highland Park included parks and grounds maintenance.
>
> In thirty years of consulting with state and local governments, Steve has performed over 160 management studies, including the 2003–2004 Organization and Staffing study and 2012 Public Works study for the city, as well as most of the studies referenced in the cover letter and in the Summary of Qualifications chapter of the proposal. He is a specialist in strategic planning; service delivery alternatives/shared services; administrative and financial services; and public works, utilities, park and recreation, fleet, and maintenance operations.

CLIENT PARTICIPATION

Clients always have some role in completing the engagement, even if only to direct the work of your firm through steering committee meetings and reviews of draft reports/deliverables. The client's role, however, may be much greater and could include client staff imbedded in your firm's engagement team. For example:

> We expect (name of client) staff will be available to assist our engagement team in the following roles:
>
> * Serve on the engagement steering committee (define who the members might be)
> * Serve as engagement liaison to help with scheduling meetings, collecting data, and facilitating engagement logistics
> * Participate on the engagement team as a financial analyst to help collect existing data and reports, review and enhance new data, and structure financial analyses to meet ongoing client needs
> * Participate in fact-finding meetings, facilitate site visits, and provide data and reports
> * Review draft deliverables
> * Participate in the presentation of the final deliverable to (name of client) leadership

ENGAGEMENT COSTS

Some clients turn right to the cost page of the proposal; others think of the cost of professional services more as a value-added issue and take a whole-proposal approach to picking a firm. In any case, this section typically is your only chance to define what you'll be paid, although negotiations may continue when the most technically competent firm is selected in a two-step procurement (qualifications, then price).

The cost section will differ, based on the nature of your professional service. Do you deliver something like a product (reasonably consistent from one client to another and definable) or a service (variable from one client to another)?

Product Oriented

A residential lot boundary survey and a home inspection are examples of professional services that are relatively consistent client-to-client or within client categories (e.g., commercial, residential). These prices are fairly repetitive client-to-client, unless something unusual is presented (e.g., a survey of a large estate or piece of land).

The cost/price challenge is to analyze what resources and staff time are required to serve the typical client (or several categories of typical clients). Your calculation could be based on the cost of a product line (like the annual budget for the home inspections division) divided by the number of inspections conducted per year, or on the cost of a single inspection based on a detailed process and cost analysis. Based on these analyses, a standard residential lot survey might cost $350; a standard home inspection $350–$400 for an average-sized home; and $500–$600 for a very large estate home.

Service Oriented

Analytical services (such as a management consulting study, design of a building by an architect, or design of a road by a consulting engineer) will vary from client to client and from project to project, as the scope and size of the project changes. In these situations, the key challenge is to define the hours required to complete the engagement and the associated expenses to be incurred (e.g., travel, report production, office/technical support).

Standard Billing Rates

The first step in preparing a cost proposal for a service is to develop standard billing rates, or a range of standard rates, for various classifications of professional staff. These hourly or daily rates must fully recover *all* costs that are not directly billed as expenses. Personnel costs, such as salaries and benefits, tend to make up a heavy percentage of the standard rate. Other cost factors include management (people not billing time to the project but supporting the firm in general ways, like the president/ CEO), administrative and support services (accounting, human resources, and information technology), rent, utilities, supplies, and training.

Service-oriented rates are calculated based on estimated billable hours, not paid hours (like 2,080 per year). A staff lawyer, for example, may be expected to bill 2,000 to 2,500 hours a year, but a law firm partner only 1,700. The cost base for each position category would be determined, and then divided by the projected annual billed hours to determine an hourly billing rate.

In Steve's management consulting practice, he varies his billing rate by the client's size (smaller communities may be billed a lower rate) and the complexity of the project (policy studies and litigation support are billed at the highest rate). Frank's billing rate also may vary from client to client, partly based on his role on the project and on the needs of the client. Barry has a single standard billing rate.

Expenses

The second step is to determine if actual expenses are to be billed, if per diem rates are billed, or if some combination of actual and per diem are billed (e.g., actual air fare plus per diem for hotel and meals). In addition, some clients have internal rules limiting expenses that also apply to professional service firms (e.g., a meal limit of $50 per day).

Administrative Matters

The Engagement Cost section of the proposal often is the place to discuss administrative requirements, such as your proposed billing protocol, compliance with any special legal requirements, and confirmation of insurance coverage (liability, auto, workers compensation), as well as to insert any forms or certifications required by the RFP (see Table A.1 for examples of certifications and agreements appended to a law firm's proposal of service).

CLOSING THE PROPOSAL

The Conclusion section consists of a short closing statement that typically includes a thank-you for the opportunity to propose. It also should provide information on how to reach the appropriate people in the firm if there are questions or additional information is required. For example:

> We appreciate the opportunity to serve (client name) on this important project. We are confident our proposal offers significant benefits at a competitive price. If you have any questions or require additional information, please contact (name of contact person) at (phone number and e-mail address).

ATTACHMENTS OR APPENDIXES

Proposals should be as short as possible, while completely communicating the proposal elements described above. In some cases, you may want to append additional information, which could include the following:

- Detailed résumés for proposed project staff
- Firm brochures and marketing materials
- Examples of work for prior clients, which could be entire reports, extracts of reports, or segments of deliverables, such as architectural and engineering firms that present a portfolio of work on comparable projects

PROPOSAL PRESENTATION

Although it is possible for a professional service to be sole sourced, a prospective client often will ask qualified professional service firms to propose and then will meet with each proposer or several finalists to provide the opportunity for discussion and defense of their proposals. The proposal presentation is your last opportunity to make a big impact on the client's decision, so it's really important. Due to strict procurement rules in the public sector, the presentation may be the first time you actually meet with key client staff, and not just Purchasing staff, to discuss the engagement opportunity. In the private sector, prior face-to-face discussions likely have occurred before the proposal is issued.

In the thirty to sixty minutes you have for your presentation, you will want to review the proposal *and* present something unique that amplifies what's written in your proposal, like a personal (and hopefully heart-felt) discussion of

- Your firm's vision for services to its clients
- Who you are, professionally and personally
- What experiences, skills, and intangibles you bring to the table
- Your understanding of why the client needs this service and how it will benefit them
- How your vision and personnel connect to this opportunity to serve the client

A key decision is who should make the presentation for your firm and how to split up the various sections in the presentation, if two or more firm personnel attend.

Unless the project is small, short in time, and straightforward in delivery (like the survey and inspection products discussed in the prior section), we suggest a two-person presentation team with these assignments:

- Relationship leader or engagement director to introduce the firm and discuss the "heart-felt" topics above
- Project manager to review the *how* (Approach and Work Plan), *when* (Schedule), *what* (Objectives and Scope of Work), and *whom* (Project Team) parts of the proposal—the nuts and bolts of conducting the engagement, with a strong focus on process and the nature of the deliverables
- Relationship leader or engagement director to review the price and wrap up the presentation

The presentation should be tailored to your audience and setting. A formal board meeting likely will require a PowerPoint or similar presentation, while an informal meeting with the management team may require only something more informal, such as a handout of key talking points.

POST-PROPOSAL FOLLOW-UP

The day after a presentation you should write (if time until the decision permits) and/or e-mail a thank-you note to key participants in the presentation, as well as to key decision makers who may not have attended your presentation. This follow-up note might include supplemental information or answers to questions raised at your presentation, including more extensive responses to questions or topics that you might not have handled as well as you would have liked during the presentation.

POST-DECISION REVIEW

Win or lose, several tasks must be completed after the client selects a professional service firm:

- *If you win*, call and thank them profusely, then get going on a contract. If you don't know by now, find out who the bidders were and what the price was for each firm's bid.
- *If you lose*, you should thank the client for the opportunity to propose and wish them well with the firm they engaged. In addition, ask them what they perceive as your firm's and your proposal's strengths and weaknesses, and how the selected firm and its proposal were superior. Finally, make sure you know who bid and what price they proposed, to determine if you were price competitive.

Both the winning and losing review may cause you to substantially alter the content or structure of your standard proposal to make it more competitive in the future. And, especially in the public sector, the review provides a look inside your competition's services, experience, and pricing strategy.

SAMPLE PROPOSAL OUTLINES FOR LEGAL AND CONSULTING SERVICES

Tables A.1 and A.2 that follow provide the table of contents for a legal proposal to provide bond counsel services and a management consulting proposal for a Parks and Recreation Department study.

TABLE A.1

Structure of a Law Firm's Proposal for Bond Counsel Services to a County

Title Page
Cover Letter
  1. General Information about the Firm
  2. Qualifications of the Firm
  3. Qualifications of Assigned Staff
  4. Approach to Services
  5. Client References
  6. Record of the Firm
  7. Liability Insurance Coverage
  8. Additional Information about the Firm's State and National Public Finance Practice
Appendix A: Cost Proposal
Appendix B: Contractor Affidavit and Agreement
Appendix C: Immigration Compliance Certificate
Appendix D: Addenda Acknowledgment
Appendix E: Professional Liability Insurance Certification

TABLE A.2
**Structure of a Management Consulting Firm's Proposal
for a Parks and Recreation Department Study**

Title Page
Cover Letter (with Table of Contents)
   I. Understanding of the Engagement
      A. Current Situation
      B. Engagement Overview (Need, Purpose, Scope, Issues)
      C. Engagement Schedule (Overview)
      D. Engagement Deliverables
  II. Summary of Our Approach
      A. Experienced Engagement Team
      B. Participative Approach
      C. Structured Work Plan Using Engagement-Tested Methods and Tools
 III. Work Plan and Schedule
      A. Work Plan
      B. Schedule (Detailed)
 IV. Engagement Management and Staffing
      A. Engagement Team
      B. Client Responsibilities
  V. Summary of Qualifications
      A. Introduction to the Firm
      B. Representation Engagements
      C. References
 VI. Cost Proposal (with Billing Protocol and Insurance Coverage)
Appendix A: Firm's Complete Client/Engagement List
Appendix B: Engagement Team Resumes

Appendix B: Deliverable Preparation Guide

An engagement report serves several purposes. First, it is evidence of completion of an engagement or a phase of an engagement. Second, it establishes a written record of the nature of the engagement and the results achieved. And third, it is the engagement team's expression of technical competence—both to the client and to other prospective clients. It should be a professional service firm's policy to submit a written report to the client at the end of an engagement. In addition, reports might be appropriate at key milestones or the conclusion of engagement phases.

The following discussion describes the approach to report preparation that the authors have followed for many years. It should be considered as a guide only, given that the professional service industry has such a wide variety of sectors.

TYPES OF REPORTS

There are two basic types of engagement reports: the letter report and the full technical report. The engagement report may be written by anyone on the engagement that is technically qualified, but it must be reviewed, approved, and signed by an appropriate and designated person or classification of person. For example, in many public accounting firms, only a partner in the firm may sign a report—and it must be signed in the firm's name; in essence, the partner is claiming full responsibility for the report's contents to his/her other partners. The engagement report is delivered to the client through the use of a brief transmittal letter, which also is signed by a properly designated person (using that person's signature).

LETTER REPORT

The letter report typically is used to document, in relatively short form, the findings, conclusions, and recommendations of the engagement. It may include reference to an accompanying engagement deliverable, such as an architect's drawings or an attorney's legal documents. The letter report format is often used when working closely with a client executive who is not interested in a lengthy report but wants brief documentation of what the engagement has accomplished.

As a specific example, a particular letter report might contain the following items of information:

- Salutation
- Background and purpose (objectives and scope) of the engagement
- Key findings
- Conclusions

- Recommendations (including reference to accompanying deliverables, if any)
- Client benefits
- Closing paragraph (including appropriate thank-yous)
- Firm's signature

Again, the letter report format is used in appropriate circumstances, particularly when accompanying deliverables cannot be incorporated in a full technical report.

FULL TECHNICAL REPORT

A full technical report typically consists of two parts: an introductory letter and an attached technical report. Management consultants, for example, generally use this format, as the primary product of their work is a technical report—as well as a presentation of their findings, conclusions, and recommendations.

INTRODUCTORY LETTER

The introductory letter contains similar information items to the *letter report* described earlier (i.e., it is a short and crisp version of the attached technical report). The introductory letter is designed to fully inform client executives of the purpose of the engagement, as well as the key findings, conclusions, and recommendations arising from the engagement. A client executive may not wish to dig into the attached technical report, but it is there if needed. As a result, the introductory letter needs to be written in an engaging manner that will motivate the reader to implement the recommendations. A key to that motivation is a clear understanding of the client benefits that should be derived from the engagement.

TECHNICAL REPORT

The technical report may take any form that meets the client's needs but should fully discuss the purpose, findings, conclusions, recommendations, and client benefits of the engagement. It follows the introductory letter, with the first page containing at least the engagement title, the issue date, the client's name, and the firm's name. Next comes the table of contents, which is designed to help the reader find a particular topic in the report and see how the report is organized. Of course, the table of contents is prepared after the technical report has been finished. It lists all of the section headings and subheadings that appear in the body of the report, along with their respective page numbers. The table of contents also includes any illustrations, exhibits, and appendixes that appear in the technical report. Then comes the body of the report.

TRANSMITTAL LETTER

The transmittal letter should be no more than one page, referencing the attached letter report or the full technical report. It is placed on the firm's letterhead and addressed to a specific client executive or group of executives (e.g., board of directors). The body of the transmittal letter might say: *We are pleased to submit our report to (name of client) that summarizes financial and retirement issues that need to be addressed, with a road map on how to address them. We also provide a written Succession Plan for each position.*

A properly designated person, using a personal signature, signs the transmittal letter.

PREPARING THE TECHNICAL REPORT

Drafting of the technical report begins with a detailed outline, similar to the final table of contents. The outline should be prepared to reflect at least all of the sections, major side headings, and minor side headings that are planned for the report. As was mentioned earlier, any or all of the engagement team members may participate in outlining and writing the technical report; however, it is important that each element of work be reviewed and approved by the lead person on the firm's engagement. That lead person should know the following about the key potential readers of the full technical report, such as follows:

- Organization level
- Expectations of outcomes of the engagement
- Familiarity with the material being discussed
- Attitudes
- Interest in the subject

These and other factors may need to be taken into account when drafting the technical report. One of the questions that needs to be asked is, "Will the client see what has been reported as criticism of their management or operation?" If so, then this issue must be addressed head on and resolved.

Generally, the opening section of the technical report should reflect elements of the service proposal or contract, as well as amendments thereto. This would include the background, objectives, scope, approach, and deliverables. Subsequent sections, for example, might include:

- Findings and observations
- Conclusions
- Recommendations
- Conceptual or detailed designs, if appropriate
- Client benefits

Given the breadth of the professional service industry, it is not prudent to specify the contents and makeup of a particular report. However the above listing of typical sections may serve as guidance.

INTERNAL DISCIPLINE AND CONTROL CONSIDERATIONS

Many professional service firms provide documentation and training regarding how they want, if not require, reports to be written. This may get down to the specific formats to be used, as well as norms for punctuation and tips for improving the report's readability. At a minimum, each firm should prepare a report that could be referred to and modeled by a report writer. Documentation and training for how reports should be written will allow the firm to gain greater consistency and provide a better service to the client.

Index